Prepper's Surviva

# 20 BOOKS IN 1

## Long-Term Resilience with Life-Saving Tactics, Stockpiling Wisdom, Water Filtration, Off-Grid Living, and Self-Defense

**Austin Cross**

© Copyright 2023 by Austin Cross - All rights reserved.

This content is provided with the sole purpose of providing relevant information on a specific topic for which every reasonable effort has been made to ensure that it is both accurate and reasonable. Nevertheless, by purchasing this content you consent to the fact that the author, as well as the publisher, are in no way experts on the topics contained herein, regardless of any claims as such that may be made within. As such, any suggestions or recommendations that are made within are done so purely for entertainment value. It is recommended that you always consult a professional prior to undertaking any of the advice or techniques discussed within.

This is a legally binding declaration that is considered both valid and fair by both the Committee of Publishers Association and the American Bar Association and should be considered as legally binding within the United States.

The reproduction, transmission, and duplication of any of the content found herein, including any specific or extended information will be done as an illegal act regardless of the end form the information ultimately takes. This includes copied versions of the work both physical, digital and audio unless express consent of the Publisher is provided beforehand. Any additional rights reserved.

Furthermore, the information that can be found within the pages described forthwith shall be considered both accurate and truthful when it comes to the recounting of facts. As such, any use, correct or incorrect, of the provided information will render the Publisher free of responsibility as to the actions taken outside of their direct purview. Regardless, there are zero scenarios where the original author or the Publisher can be deemed liable in any fashion for **any damages or hardships that may result from any of the information discussed herein.**

Additionally, the information in the following pages is intended only for informational purposes and should thus be thought of as universal. As befitting its nature, it is presented without assurance regarding its prolonged validity or interim quality. Trademarks that are mentioned are done without written consent and can in no way be considered an endorsement from the trademark holder

# Contents

**INTRODUCTION** .................................................................................................................. 13

**BOOK 1: PREPPER'S FOUNDATION: ESSENTIAL READ** ........................................................ 15

    Emergency Planning and Preparedness .................................................................... 15

    Food Storage and Preservation Techniques ............................................................. 15

    Procuring, Purifying and Storing Water .................................................................... 15

    First Aid and Medical Preparedness ......................................................................... 16

    Shelter and Home Security Measures ...................................................................... 16

    Emergency Communication Strategies ..................................................................... 16

    Essential Gear and Tools for Survival ........................................................................ 16

    Self Defense and Personal Protection ...................................................................... 17

    Navigation and Mapping Skills .................................................................................. 17

    Community Building and Mutual Aid Networks ....................................................... 17

    Emergency Evacuation Planning ............................................................................... 18

    DIY Project: Emergency Go-Bag Assembly ............................................................... 18

**BOOK 2: PREPARING FOR ANY SCENARIO** ........................................................................ 20

    Forming an Emergency Plan ..................................................................................... 20

    Stockpiling Essential Supplies of Food, Water and Medicine .................................. 20

    Establishing Skills to Foster Independence .............................................................. 20

    Building Resilience: Physical and Mental Preparedness .......................................... 20

    Establish Communication Protocols ......................................................................... 21

    Securing Shelter and Secure Spaces ......................................................................... 21

    Understanding and Utilizing Renewable Energy Sources ........................................ 21

    Discovering Basic Survival Techniques ..................................................................... 21

    Establishing Community Networks and Support Systems ....................................... 21

    Preserving Financial Stability during Uncertain Times ............................................ 22

    Practice of Permaculture and Sustainable Living ..................................................... 22

    Enhancing DIY and Repair Skills ................................................................................ 22

    Implement Home Security Measures ....................................................................... 22

    Develop and Stockpile an Emergence Plan .............................................................. 22

Understanding Psychological Responses to Crisis .................................................................. 23

Integrating Technology for Monitoring and Communication .................................................... 23

Conduct Regular Drills and Training Exercise ......................................................................... 23

Adaptation and Thriving in Shifting Environments .................................................................. 23

DIY Project: Emergency Go-Bag Assembly ............................................................................. 23

**BOOK 3: PREPAREDNESS HANDBOOK: CRISIS READY** ............................................................ 26

Customized Crisis Response Plans ......................................................................................... 26

Building Resilient Supply Chains ............................................................................................. 26

Implementation of Communication Strategies ........................................................................ 26

Establish Emergency Contact Protocols .................................................................................. 26

Drills and Exercise for Crisis Regular Conducting ................................................................... 27

Emergency Employee Response Training ............................................................................... 27

Maintain Continuity Business Plans ........................................................................................ 27

DIY Project: Emergency Communication Plan Creation ......................................................... 27

**BOOK 4: CRISIS MIND: NAVIGATING MENTAL HEALTH** .......................................................... 30

Recognizing Signs of Psychological Distress .......................................................................... 30

Coping Strategies to Manage Stress and Anxiety ................................................................... 30

Seek Help: Resources and Hotlines ........................................................................................ 30

Building Resilience with Positive Psychology .......................................................................... 30

Self-Care Techniques to Achie Mental Well-being .................................................................. 30

Supporting Others in Crisis ...................................................................................................... 31

Attaining Independence by Overcoming Stigma ..................................................................... 31

Create a Safe and Supportive Environment ............................................................................ 31

Mindfulness and Meditation Techniques ................................................................................. 31

Manage Trauma and Post-Traumatic Stress Disorder (PTSD) ............................................... 31

Addressing Substance Abuse and Addiction Issues ............................................................... 32

Balancing Work and Personal Life during Crisis ..................................................................... 32

Understanding the Impact of Crisis on Children and Adolescents .......................................... 32

Cultivating Healthy Relationships and Boundaries ................................................................. 32

Exploring Alternative Therapies and Healing Modalities ......................................................... 32

DIY Project: Mind Matters: Navigating Mental Health in Crisis ............................................... 33

## BOOK 5: WATER MASTERY: SURVIVAL TECHNIQUES ........ 35
Recognizing Safe Water Sources in the Wild ........ 35
Methods for collecting and storing water. ........ 35
Techniques for Water Filtration and Purification ........ 35
Building DIY Water Filtration Systems ........ 35
Utilizing Natural Water Purification Techniques ........ 36
Emergency water treatment chemicals and tablets ........ 36
Solar Water Disinfection System ........ 36
Distillation: Purifying Water Through Evaporation ........ 36
Understanding Waterborne diseases and prevention ........ 37
Calculating Water requirements for survival ........ 37
Hydration Strategies in Extreme Environments ........ 37
DIY Project: Water Worries: Mastering Survival Hydration Techniques ........ 37

## BOOK 6: PANTRY PREP: FOOD SECURITY MASTERY ........ 40
Understanding Food Security: Exploring Its Significance ........ 40
Establish a Comprehensive Food Storage Strategy ........ 40
Stockpiling Long-term Food Staples ........ 40
Establish an Emergency Food Pantry ........ 40
Rotating and Managing Food Inventory ........ 41
Proper Packaging and Storage Techniques for Food Products ........ 41
Utilize Freeze-Dried and Dehydrated Foods ........ 41
Canning and Preserving Fresh Produce ........ 41
Understanding Expiration Dates and Shelf Lives ........ 42
Establish a Garden to Harvest Fresh Vegetables ........ 42
Acquisition of Hunting and Foraging Skills ........ 42
Implementation of Permaculture Practices ........ 42
Exploring Alternative Protein Sources ........ 43
Address Dietary Restrictions and Allergies in Emergency Planning ........ 43
Make Emergency Meal Plans ........ 43
Utilizing Community Resources for Food Security ........ 43
Bartering for Food in Crisis Situations ........ 43

Emergency Cooking Methods Without Electricity ............ 44

Continued Update and Revamp of Food Security Plan ............ 44

DIY Project: Food Security: Preparing Your Pantry for Any Eventuality ............ 44

**BOOK 7: PREPPER'S RECIPE COLLECTION: SURVIVAL CUISINE** ............ 47

Vegetable Soup ............ 47

Pasta Primavera ............ 47

Tuna Casserole ............ 47

Black Bean Quesadillas ............ 48

Clean-up Convenience with Minimal Meals ............ 48

Benefits of One-Pot and One-Pan Meals ............ 48

Example One-Pot and One-Pan Meals ............ 48

Tips for Easy Cleanup ............ 49

Classic Chicken Noodle Soup ............ 49

**Tomato Basil Soup** ............ 49

Bean and Pasta Soup ............ 50

Minestrone Soup ............ 50

Sustained Energy Mixes for Trail Running and Bars DIY ............ 50

Meal Ideas for Outdoor and Campfire Cooking ............ 50

Shelf-Stable Ingredients for Sweets and Desserts ............ 51

Creative Recipes Utilizing Foraged and Wild Edibles ............ 51

Sauerkraut ............ 51

Pickled Cucumbers ............ 51

Pickled Jalapenos ............ 52

Protein-Rich and Eco-Friendly Dishes ............ 52

Baking with Homemade Legumes and Lentils ............ 52

Cooking Conservation Strategies: ............ 52

Pantry Essentials to Prepare International Cuisine ............ 53

Comfort Foods for Stressed Times ............ 53

DIY Project: Survival Cuisine: The Prepper's Recipe Collection ............ 53

**BOOK 8: FIRE MASTERY: IGNITE SURVIVAL ART** ............ 56

Principles for Lighting Fires Easily ............ 56

Essential Tools and Equipment for Fire-Making .................................................................. 56

Items Essential to Assembling a Fire-Making Kit ................................................................ 57

Materials and Kindling ........................................................................................................ 57

Lays: Teepee, Lean-to, etc. Cabin/Log ................................................................................ 57

Starting Fire in Challenging Conditions ............................................................................... 58

Starting From Scratch ......................................................................................................... 58

Starting Fire DIY Resin, Pine ............................................................................................... 58

Considerations for Fire Safety ............................................................................................. 58

Maintain Safe Fire-Building and Management Behavior ..................................................... 59

Understanding Open Flame Cooking Techniques ............................................................... 59

Rescue using Fire for Signaling ........................................................................................... 60

Extending and Maintaining Fire Safety ................................................................................ 61

Practice Makes Perfect: Building Fire Proficiency ............................................................... 61

DIY Project: Ignite Survival: Mastering the Art of Fire Making ............................................. 62

## BOOK 9: PREPPER'S HERBAL MEDICINE HANDBOOK ................................................ 64

Benefits and Limitations of Herbal Remedies ..................................................................... 64

Prepper's Herbal Medicine Cabinet Construction ................................................................ 64

Identification and Harvest of Wild Medicinal Plants ............................................................. 64

Herbal First Aid for Common Injuries and Ailments ............................................................. 65

Herbal Antiseptics and Wound Care Solutions .................................................................... 66

Herbal Remedies for Pain and Inflammation Management ................................................. 66

Immune-Boosters and Herbal Preventative Solutions ......................................................... 67

Herbal Remedies for Respiratory Illnesses and Allergies .................................................... 67

DIY Project: Herbal Healing Salve ....................................................................................... 68

## BOOK 10: SELF-SUFFICIENT LIVING: OFF-GRID EMBRACE ......................................... 70

Location Selection for Off-Grid Living .................................................................................. 70

Sustainable Homes: Eco-Friendly Designs for Off-Grid Living ............................................ 71

Off-Grid Food Production with Gardening and Permaculture .............................................. 71

Waste Management in Off-Grid Settings ............................................................................. 72

Transportation Solutions Off-Grid ........................................................................................ 72

Financial Planning for Off-Grid Living .................................................................................. 72

Health Medical Care Off-Grid: Self-Sufficient Practices........................................................... 73

DIY Project: Off-Grid Solar Power System............................................................................ 74

## BOOK 11: HOME DEFENSE FOR PREPPERS ........................................................................... 76

Home Security Systems and Surveillance Technology......................................................... 76

Establish a Strong Perimeter with Fences, Gates, and Barriers............................................ 76

Establishing Safe Rooms and Emergency Shelters............................................................... 76

Home Defense Weapons and Training ................................................................................. 77

Establishing Neighborhood Watch Programs ...................................................................... 77

Utilizing Guard Dogs for Home Protection .......................................................................... 77

Develop Code Words and Signaling Systems ....................................................................... 78

Implementing Home Security Drills and Exercise ................................................................ 78

Maintaining Operational Security ........................................................................................ 79

Understanding Threat Levels and Response Protocols........................................................ 79

DIY Project: Home Fortress: Strengthening Your Defenses ................................................. 80

## BOOK 12: CAMPING WITH RVS ............................................................................................ 82

Setting Up an RV for Off-Grid Living ..................................................................................... 82

Management of Water in an RV ........................................................................................... 82

Solar Panels and Generators ................................................................................................ 82

Securing and Strengthening Your RV to Prevent Threats .................................................... 82

Mobile Communication Solutions for Remote Areas .......................................................... 83

Stockpiling Essential Supplies in Limited Space ................................................................... 84

Cooking and Food Prep in an RV Kitchen............................................................................. 84

Wilderness Survival Skills for RVers ...................................................................................... 85

Waste and Sanitation in an RV ............................................................................................. 86

Waste and sanitation management..................................................................................... 86

Security Measures for RVs and Motorhomes ...................................................................... 87

RV and motorhome security ................................................................................................ 87

DIY Project: RV Survival Kit: Preparing Your Mobile Shelter ................................................ 87

## BOOK 13: PREPPER'S COMMUNITY: SHARING PREPAREDNESS ............................................ 90

Analyzing Community Strengths and Weaknesses .............................................................. 90

Emergency Response Teams................................................................................................ 90

Establishing Neighborhood Watch Programs ..................................................................................... 90

Preparedness Workshops and Training Hosted ................................................................................. 91

Develop Communications Lists and Networks .................................................................................. 91

Establishing Relationships with Local Agencies ................................................................................. 92

Organization of Aid Networks and Mutual Aid Organizing ................................................................ 92

Conduct Risk Evaluations .................................................................................................................. 92

Establishing Emergency Shelters and Spaces .................................................................................... 93

Promoting Diversity and Inclusion in Preparedness Efforts ............................................................... 93

Strengthening Social Cohesion Through Shared Experiences ........................................................... 93

Implementing Sustainable Practices for Resilience ........................................................................... 94

Supporting Vulnerable Populations in the Community ..................................................................... 94

Fostering a Culture of Preparedness Through Proactive Engagement .............................................. 95

Leveraging Community Resources for Disaster Recover ................................................................... 95

Reconciling Resilience Efforts: Recognizing and Praising Resilience Strategies ................................. 95

Advocating for Policy Changes that Support Community Resilience ................................................. 95

Integrating Technology and Alerts for Emergency Communication .................................................. 96

Continuous Evaluation and Enhancing Resilience Plans .................................................................... 96

DIY Project: Community Emergency Preparedness Plan: Strengthening Bonds in Crisis .................. 96

## BOOK 14: ENERGY CREATION IN SURVIVAL SCENARIOS ................................................................. 99

Assess Energy Needs and Scenarios .................:............................................................................... 99

Utilizing Renewable Energy Sources ................................................................................................. 99

Portable Energy Solutions ................................................................................................................. 99

DIY Energy Generation ...................................................................................................................... 99

Batteries Vs Capacitors ................................................................................................................... 100

Improvising Emergency Energy Sources in Critical Situations ......................................................... 100

Prioritizing Energy Usage to Achieve Survival in Critical Scenarios ................................................. 102

DIY Project: Portable Solar Generator: A Versatile Energy Solution in Survival Scenarios ............. 102

## BOOK 15: SURVIVAL SKILLS IN THE WILD ....................................................................................... 104

Fire Starting Techniques: Flint Steel and Bow Plow ........................................................................ 105

Navigation Skills Are Vital to Survival .............................................................................................. 105

Wilderness First Aid: Treating Injuries, Ailments ............................................................................ 106

Hunting and Trapping Techniques and Tools ............................................................................... 107

Fishing: Tactics for Varying Environments ................................................................................... 107

Building and Utilizing Primitive Tools: Spears and Arrows ........................................................ 107

Weather Prediction and Animal Behavior .................................................................................... 107

DIY Project: Wilderness Survival Kit: Essential Gear for Outdoor Adventures ....................... 107

## BOOK 16: COMMUNICATION TACTICS FOR CRISIS .................................................................. 110

Emergency Equipment: Radios and Phones ............................................................................... 110

Code Words and Signals to Establish Secure Communication ................................................ 110

Setting Up Neighborhood Networks to Facilitate Communication .......................................... 110

Utilizing Social Media for Crisis Communication ....................................................................... 110

Emergency Communications with Mobile Alerts, Updates and Apps ..................................... 111

Contact Lists of Family, Friends, and Essential Services .......................................................... 111

Licensing and Operating Ham Radio ............................................................................................ 112

**Communication With Local Authorities: Establishing Contact** ............................................. 112

DIY Project: Emergency Communication Kit: Staying Connected During Crisis ................... 114

## BOOK 17: THE PREPPER'S AID GUIDE ........................................................................................ 116

Basic Anatomy and Physiology for First Aid ............................................................................... 116

CPR (Cardiopulmonary Resuscitation) and AED Use ................................................................ 116

Treating Severe Bleeding and Wound Management ................................................................. 116

Splinting Fractures and Immobilizing Injuries ............................................................................ 116

Acknowledging and Addressing Shock ....................................................................................... 117

Approaches for Treating Burns at Different Degrees ................................................................ 117

Coughing .......................................................................................................................................... 117

Treating Head Injuries and Concussions .................................................................................... 117

Allergic Reactions and Anaphylaxis ............................................................................................. 117

DIY Project: Prepper's Comprehensive First Aid Kit: Essential Supplies for Emergency Aid ...... 118

## BOOK 18: CITY SURVIVAL TACTICS ............................................................................................ 121

Building Urban Navigation Skills with Maps, GPS, and Landmarks ........................................ 121

Building Emergency Kits for Urban Survival .............................................................................. 121

Urban Environments for Securing Shelter .................................................................................. 121

Locating and Purifying Water in Urban Settings ........................................................................ 121

Gathering Food and Supplies in Urban Areas.................................................................................................122

Utilizing Public Transportation and Alternative Routes ...................................................................................122

Navigating Urban Traffic and Roadblocks .......................................................................................................122

Securing Your Apartment or Home in an Urban Crisis .....................................................................................122

Maintaining Communication in Urban Areas ...................................................................................................122

Avoid Crowded and Dangerous Areas ..............................................................................................................123

DIY Project: Urban Survival Kit: Navigating City Environments in Crisis ......................................................123

**BOOK 19: SHELTER ESSENTIALS: MAINTAINING CLEANLINESS AND SAFETY** ...............................126

Ventilation to Prevent Mold and Mildew ..........................................................................................................126

Precautions to Avoid Fire Emergencies ............................................................................................................126

Identification of Structural Weaknesses and Damage .......................................................................................126

Installing Carbon Monoxide Detectors in Living Spaces ..................................................................................127

Reducing Accidental Incidents When Storing Hazardous Items .......................................................................127

Clear Away Clutter for Safe Paths ....................................................................................................................127

Proper Storage of Hazardous Chemicals and Materials ....................................................................................127

Regular Testing of Systems, Alarms and Generators .......................................................................................127

Establish Emergency Communication Protocols ..............................................................................................128

Implement Security Measures to Prevent Break-Ins ........................................................................................128

Training Residents on Safety Procedures .........................................................................................................128

Lighting an Entire Shelter .................................................................................................................................128

Maintaining Stockpiling Kits of Food ..............................................................................................................128

Maintain Hygiene and Sanitation Facilities ......................................................................................................129

DIY Project: Shelter Maintenance Kit: Keeping Clean and Safe in Emergency Shelters.................................129

**BOOK 20: GARDENING FOR PREPPERS: ESSENTIAL GUIDE** ................................................................131

Understanding Soil Composition and pH Levels ..............................................................................................131

Selecting Appropriate Plants Based on Climate and Soil Conditions ...............................................................131

Planning Your Garden Layout to Achie Optimum Yield ..................................................................................131

Companion Planting ..........................................................................................................................................131

Seed Saving and Preservation Techniques ........................................................................................................131

Building Raised Beds or Container Gardens for Limited Space .......................................................................132

Watering Techniques to Achie Optimal Plant Health .......................................................................................132

Fertilization Options Including Organic and Synthetic Solutions ................................................................. 132

Survival Gardens Provide an Ideal Platform ........................................................................................... 132

Season Extension Solutions ................................................................................................................... 132

Vertical Gardening for Space Efficiency ................................................................................................. 133

Permaculture Principles for Sustainable Gardening .............................................................................. 133

Mulching and Weed Control Strategies ................................................................................................. 133

Integrating Medicinal Herbs and Edible Flowers into Your Garden ....................................................... 134

DIY Project: Survival Garden: Growing Your Green Thumb for Self-Sufficiency ..................................... 137

**CONCLUSION** ............................................................................................................................................ 139

BONUS ....................................................................................................................................................... 140

# INTRODUCTION

Welcome to the Prepper's Survival Guide: 20-in-1! In today's uncertain world, being prepared for anything can only serve to increase one's chance of surviving adverse conditions and can make all the difference when facing natural disasters, economic instability, or social disruptions. Knowing and mastering survival techniques for such circumstances are invaluable assets in times of trouble.

This comprehensive guide serves as your roadmap towards long-term resilience, providing lifesaving tactics, stockpiling advice, water filtration techniques, off-grid living strategies and self-defense tactics based on years of experience from seasoned preppers. Armed with this information and tools provided here by this book, any crisis will be navigated without panic!

With these pages, you will gain practical advice for building an efficient survival kit, procuring essential resources, and safeguarding your home as you maintain self-sufficiency despite adverse circumstances. From food preservation techniques and renewable energy solutions to emergency medical care and wilderness survival skills - each chapter delivers actionable insights that help prepare you for whatever may come.

No matter your prepper level or experience level, The Prepper's Survival Guide: 20-in-1 will serve as your essential companion in protecting yourself and loved ones against an unpredictable world. So, let's embark together on this journey, arming ourselves with knowledge necessary for survival in any circumstance.

This comprehensive guide serves as your ultimate companion in long-term preparedness and survival. Drawing upon decades of expertise from veteran preppers and survival specialists, this book distills their collective wisdom into one easily consumable resource. Perfect whether you are starting from scratch as a prepper or looking to add new tools into your arsenal; whether novice or expert alike can all find value here!

Within these pages you'll discover an unparalleled repository of knowledge spanning across an extensive list of essential subjects, spanning emergency preparedness to advanced survival tactics. Each chapter's goal is to equip you with skills and strategies needed for succeeding no matter the circumstance; whether dealing with short-term crises or anticipating longer term uncertainty. With our insights as your guide, we hope this knowledge can help you face them with confidence and competence!

Topics covered will include creating an emergency survival kit suited to your own specific needs; stockpiling food, water, and essential supplies in preparation of emergencies; as well as mastering water purification techniques.

Establishing off-grid living arrangements to achieve sustained self-sufficiency. Implement stringent security measures to protect both yourself and those closest to you, while learning essential first aid and medical skills that could come in handy during emergencies.

Navigating wilderness survival scenarios gracefully and confidently Each chapter in The Prepper's Survival Guide: 20-in-1 offers practical advice, step-by-step instructions, and real-world examples designed to put theory into action. Whether hunkering down at home or venturing out into nature - The Prepper's Survival Guide has your covered!

Are you ready to take control of your future and embrace self-reliance and resilience as principles to guide this journey together? If that's the case, let's embark upon this adventure together! Your journey ahead may present many obstacles, but with knowledge and preparation you can tackle whatever comes your way with courage and dignity. Your journey towards long-term resilience begins here! This comprehensive guide serves as your essential companion on this path towards long-term preparedness and survival. Drawing upon the collective knowledge and expertise of veteran preppers and survival experts, this book distills decades of knowledge into one invaluable source. No matter your level of prepper or survival expertise - from novice seeking solid grounding for preparation or expansion into more areas - there's something here for everyone in this book!

Within these pages lies an immense repository of knowledge covering an expansive spectrum of essential topics - from emergency preparedness basics to advanced survival tactics - designed to equip you with all of the skills and strategies required for survival in any circumstance - be it short term crisis response or planning ahead for more long term uncertainty - the insights contained within this guide can help you navigate any difficulties with ease and competence.

# BOOK 1: PREPPER'S FOUNDATION: ESSENTIAL READ

Understanding and assessing risks are central components of making informed decisions - be they personal or professional in nature. Risk analysis involves recognizing potential hazards, assessing their probability of occurrence, and projecting their impact to help people prioritize risks based on severity and devise strategies to minimize or manage them effectively. Visual tools like risk matrices or probability-impact diagrams help visualize this process for easier management - by thoroughly comprehending risks we can make informed choices to reduce negative outcomes while expanding opportunities for success.

## Emergency Planning and Preparedness

Emergency planning and preparedness involve developing strategies and protocols for quickly responding to crises or disasters of all kinds - be they natural (quakes, hurricanes, or flooding), human-caused such as fires or terrorist attacks or both. Effective emergency planning requires identifying potential threats, creating communication channels, assigning roles and responsibilities, and developing evacuation or shelter-in-place procedures. Preparedness efforts may include stockpiling emergency supplies, conducting drills or simulations, and informing community members on emergency protocols. By investing in comprehensive emergency planning and preparation efforts, individuals and communities can minimize loss of life or property damage during times of crises.

## Food Storage and Preservation Techniques

Food preservation techniques are indispensable skills that ensure food security during times of emergency or scarcity. Achieving this requires protecting food from spoilage, contamination, and pest infestation to preserve quality and ensure its consumption by consumers. Food preservation techniques typically utilized include canning, freezing, drying, and pickling - each using different mechanisms to avoid the growth and reactions associated with bacteria and enzymes. Proper packaging of perishable items such as vacuum sealing or using airtight containers can extend their shelf life significantly, helping individuals reduce food waste, save money, and ensure access to nutritious meals in all circumstances. By mastering food storage and preservation techniques individuals can reduce food waste while saving both time and money while assuring access to nutritious meals in times of hardship.

## Procuring, Purifying and Storing Water

Water procurement, purification, and storage are essential aspects of emergency preparedness and self-sufficiency. Accessing clean water for drinking, cooking, sanitation needs, medical purposes or simply hygiene purposes is vitally important. Emergency situations often necessitate alternative means for procuring safe drinking water sources; such methods include collecting rainwater or tapping into natural springs for supplies; portable water filtration systems may also provide valuable hydration solutions. Purification methods such as boiling, chemical treatment or UV light treatment can effectively remove pathogens and impurities from water sources. When selecting containers to store stored water in, ensure they are airtight while also rotating regularly to keep its freshness at optimum levels. By prioritizing water sourcing, purification, and storage individuals can protect their health and well-being even in difficult environments.

## First Aid and Medical Preparedness

First aid and medical preparedness involves equipping yourself with the knowledge, skills, and supplies required for emergency healthcare in case of medical crises. First aid involves providing immediate attention for injuries or medical conditions that arise, stabilizing patients until professional medical assistance arrives, as well as training essential skills like CPR, wound care and recognizing signs of medical distress. Basic first aid classes provide training that addresses essential techniques like CPR and wound care as well as recognizing distress signs in others. As part of medical preparedness, creating an accessible first aid kit that includes essential supplies like bandages, medications and medical instruments is also key. Individuals could also consider developing more advanced medical skills or earning certification in CPR, wilderness medicine or emergency response to improve their chances of handling medical emergencies in critical situations and reduce injuries and save lives. Being prepared may save lives while mitigating injuries sustained when handling them effectively can significantly decrease injury severity levels in critical situations.

## Shelter and Home Security Measures

Shelter and home security measures are key in safeguarding individuals and their families against external threats, providing a haven during emergencies, and offering peace of mind in times of trouble. These include locking all windows and doors securely; installing locks or security systems as necessary, reinforcing vulnerable entry points to deter intruders. Establishing emergency shelters or safe rooms within your home can offer **protection from severe weather conditions, such as tornadoes or hurricanes. Homeowners may benefit from investing in backup power sources like generators** or solar panels as a safeguard during power outages, to maintain functionality during any outages. Maintaining adequate supplies of food, water, and essential items such as shelter can provide essential comfort during times of isolation or disruption. By adopting home security measures individuals can increase both their security and resilience against threats or emergencies that arise.

## Emergency Communication Strategies

Communication during emergencies is of critical importance in terms of both protecting personal safety and coordinating response efforts. Establishing reliable channels allows swift dissemination of vital information to affected parties as well as authorities involved. Utilizing multiple communication mediums such as mobile phones, radios, social media, and public announcements helps reach a wide range of target audiences. Furthermore, creating an efficient hierarchy and designating spokespersons provides clarity and consistency of message delivery. Maintaining regular updates regarding a situation's current circumstances, evacuation procedures and available resources helps ease panic and confusion among residents, while community communication networks facilitate mutual aid among neighbors. Training on emergency communication protocols and conducting drills are crucial components to being ready and responsive when crises strike. Encouraging clear, timely, and accessible communication strategies promotes efficient coordination and overall emergency preparedness.

## Essential Gear and Tools for Survival

Preparing for emergencies by stockpiling essential survival gear and tools can increase one's chances of survival and comfort in hazardous circumstances. Examples include first aid kits, multi-tools, flashlights, water purification tablets, fire starters, emergency shelters and navigation equipment. Each of these items serves a distinct purpose ranging from medical aid and food supply security to fire starters and navigation tools. Personal hygiene items, extra clothing and durable footwear all play an essential part in improving physical wellbeing and morale. Therefore, creating survival kits tailored specifically to individuals' individual needs and environmental conditions is imperative to their effectiveness. Regular maintenance and replenishment of supplies ensure readiness when they're needed, while learning the correct use of each tool through training and practice further increases its efficacy. Overall, investing in essential survival gear and tools forms an integral aspect of emergency preparedness that offers peace of mind during uncertain situations.

## Self Defense and Personal Protection

Self-defense training equips individuals with both physical and mental tools that enable them to defend themselves when faced with uncertain or dangerous situations and prioritize personal safety. Training helps prepare people with strategies for deterring threats or responding effectively when attacked. Situational awareness, verbal de-escalation techniques and physical maneuvers give individuals tools they need to assert control and safeguard themselves and others from danger. Carrying non-lethal self-defense tools such as pepper spray or personal alarm can offer peace of mind while deterring threats; however, understanding legal limitations and ethical considerations surrounding self-defense is crucial to avoid unintended harm or further escalated conflict. Promoting an avoidance and prevention mindset reduces confrontations and enhances overall safety, and investing in self-defense training or adopting proactive personal protection measures enhances confidence and resilience when encountering potentially risky situations.

## Navigation and Mapping Skills

Navigation and map reading skills are vitally important when traveling through unfamiliar terrain or during emergency situations, as they allow individuals to successfully traverse terrain and reach destinations safely and successfully. Knowledge of topographic maps, compass navigation and GPS systems allows individuals to pinpoint their position on maps, plan routes and identify potential obstacles or hazards in advance. At the same time, mastering techniques such as triangulation and dead reckoning helps increase accuracy and confidence while practicing map reading under varied environments and circumstances helps hone proficiency while building self-reliance. Knowledge of landmarks, natural features, and emergency shelters aids navigation and decision-making processes. Collaborating with others and sharing information enhances collective navigation capabilities - especially during complex or rapidly shifting scenarios. Mastery of navigational skills gives people confidence when traveling safely through urban or wilderness settings alike.

## Community Building and Mutual Aid Networks

Building strong community ties and creating mutual aid networks are vital steps toward strengthening resilience and providing collective support during emergency situations. Community building initiatives, like neighborhood associations, community centers, and social events that foster interpersonal connections and increase a sense of belonging among residents are invaluable ways to strengthen interpersonal ties and foster belongingness in society. Organization of community workshops, training sessions and disaster preparedness drills provides residents with shared knowledge and abilities that promote stronger bonds between neighbors. Establishing mutual aid networks enables neighbors to pool resources, assistance, and information during crises - further supplementing official response efforts. Leveraging technology and social media platforms also enhances community communication and coordination efforts while investing in community building initiatives can foster an atmosphere of solidarity and cooperation that strengthens resilience when facing hardship.

## Emergency Evacuation Planning

Preparing for emergency evacuations is key for swift and safe relocation during crises like natural disasters or dangerous incidents. Developing evacuation plans involves outlining evacuation routes, assembly points and transportation solutions in advance. Working closely with local authorities, emergency services providers and community organizations helps ensure comprehensive planning and coordination efforts. Conducting evacuation drills and simulations helps familiarize residents with procedures, boost readiness and increase preparedness. Further, tailoring evacuation plans specifically to vulnerable groups such as elderly individuals or those living with disabilities ensure inclusivity and accessibility for these populations. Maintaining communication channels and providing timely updates during evacuations facilitates orderly and efficient movement. Re-evaluating evacuation plans regularly as circumstances or lessons learned arise improve their effectiveness and responsiveness - prioritizing thorough evacuation planning can contribute to strengthening community safety and resilience during an emergency.

## DIY Project: Emergency Go-Bag Assembly

**Materials:**

- **Backpack or sturdy duffel bag.**
- Water bottles or water bladder
- Non-perishable food items (granola bars, canned goods, etc.)
- First aid kit
- Flashlight with extra batteries
- Multi-tool or Swiss Army knife
- Emergency blanket
- Whistle
- Portable phone charger with cables
- Personal hygiene items (toothbrush, toothpaste, wipes, etc.)
- Map of local area
- Cash (small bills and coins)

**Procedure:**

- Begin by selecting an emergency go-bag bag suitable for extended wear. Lay out all the items you plan to include to assess how much space will be necessary and to be certain everything fits snugly inside.
- Before packing the bulkiest and heaviest items such as water bottles or canned goods at the bottom, add in smaller ones such as your first aid kit and personal hygiene items to fill any empty spaces and optimize space efficiency.
- Put flashlight, whistle, and other essential items in easily accessible outer pockets so they are easily retrievable when an unexpected need arises. Utilizing straps or clips, secure an emergency blanket firmly around the outside of your bag for instantaneous access in case an immediate crisis should arise.

- Guarantee that all items are packed securely to prevent shifting during transit and the potential loss or damage of fragile objects. Also, double check to include important documents like copies of IDs, emergency contact info and any relevant medical records.
- Once the bag has been packed, perform one final check to ensure you haven't overlooked anything essential before sealing it and placing it somewhere accessible if an evacuation or emergency should arise, such as near an exit or closet.
- Maintain and update your go-bag regularly to account for changing seasons, personal needs, or emergency preparedness requirements.

# BOOK 2: PREPARING FOR ANY SCENARIO

Assessing potential threats and risks involves systematically identifying and assessing hazards that could harm individuals, communities, or organizations. Detection requires an analysis of multiple factors including geographical location, historical information, and current trends to predict potential threats and assess their likelihood and effects. By understanding these risks, stakeholders can prioritize mitigation efforts and allocate resources more effectively. Common threats include natural events like earthquakes and flooding, technological hazards like chemical spills or power outages, as well as human-created risks like terrorism or cyber-attacks. Regular evaluation is vital to adapt quickly to changing conditions and stay prepared.

## Forming an Emergency Plan

Establishing an emergency plan is crucial to effectively responding to crises and disasters. A comprehensive emergency plan outlines specific actions and procedures to be undertaken during, before, after, and before recovery from crises and disasters to reduce harm while expediting recovery - typically including elements such as risk evaluation, communication protocols, evacuation procedures and resource allocation. Working closely with all relevant stakeholders as well as regularly revising it according to lessons learned or changing circumstances further strengthens its efficacy.

## Stockpiling Essential Supplies of Food, Water and Medicine

**Stockpiling essential supplies like food, water and medicine is an integral component of emergency preparedness.** By stockpiling essentials like these ahead of a disaster strikes, individuals and their families will ensure they can thrive during times of disruption when regular resources may become limited. Recommended items may include non-perishable food items, bottled water, prescription medication, first aid supplies and hygiene items that should all be rotated regularly to maintain freshness and effectiveness.

## Establishing Skills to Foster Independence

Self-sufficiency skills provide individuals with confidence when faced with challenging circumstances and reduced reliance on external assistance. Such abilities include first aid, basic survival techniques, gardening, and home repairs - as well as communication and conflict resolution abilities that foster resilience during emergencies as well as community cohesion during times of disasters. Investment of time, effort, and education towards attaining these abilities not only foster preparedness but foster an empowerment-inspired state.

## Building Resilience: Physical and Mental Preparedness

Building resilience involves developing both physical and mental resilience to effectively deal with challenges, adapt to situations that arise unexpectedly and bounce back after obstacles have presented themselves. Physical preparation includes maintaining good health, remaining physically fit and learning basic survival skills; mentally preparedness means devising effective coping strategies, stress reduction techniques and cultivating a positive mindset - developing strong support networks is also beneficial to resilience building; these efforts allow individuals to better weather emergencies before rebounding back stronger than before, increasing overall resilience and adaptive capacity in individuals over time.

## Establish Communication Protocols

Establishing communication protocols is vital to effective coordination and information dissemination during emergencies. These protocols set the rules on how individuals and organizations will collaborate before, during, and after an incident, providing reliable exchange of data. Their main components include designated channels of communication for emergency contact lists as well as procedures for reporting or verifying data. Testing and training on communication protocols help to ensure preparedness at key moments while minimizing confusion during critical moments.

## Securing Shelter and Secure Spaces

Securing shelter and safe spaces are crucial in protecting individuals and communities from emergencies and disasters, including extreme weather events, structural damage, environmental contamination, and natural hazards such as extreme temperatures. Shelter should be structurally sound with sufficient amenities such as medical support. Furthermore, safety precautions such as emergency supplies storage space provision as well as security personnel must also be implemented for optimal effectiveness in protecting individuals during emergencies and disasters.

## Understanding and Utilizing Renewable Energy Sources

Understanding and employing renewable energy sources are integral parts of increasing resilience and sustainability during emergencies when conventional sources may become disrupted. Solar, wind, and hydroelectric power all provide sustainable alternatives that reduce dependency on fossil fuels while mitigating environmental effects. By investing in these infrastructure and technologies communities can improve energy resilience while decreasing carbon emissions while cutting long-term costs; decentralized renewable systems even ensure continued power access during such emergencies.

## Discovering Basic Survival Techniques

Learning basic survival techniques equips individuals with essential tools and knowledge for dealing with emergencies in any environment, including wilderness survival skills like shelter-building, fire making and foraging for food and water. First aid training, navigation skills and self-defense techniques also increase preparedness and resilience; regular hands-on practice of these survival techniques builds confidence and competence within individuals allowing them to respond swiftly in any emergency and protect both themselves and others in adverse circumstances.

## Establishing Community Networks and Support Systems

Establishing community networks and support systems promotes cooperation, unity and resilience among individuals and groups facing common challenges. Such networks bring together various stakeholders - residents, local organizations, businesses, and government agencies alike - who share resources, information, and expertise among one another. Furthermore, community initiatives such as neighborhood watch programs, mutual aid networks or disaster response teams serve to bind residents closer together during emergencies; by developing strong community support systems communities can enhance preparedness efforts as well as recovery and increase resilience and wellbeing overall.

## Preserving Financial Stability during Uncertain Times

Financial security during times of instability is critical to weathering economic challenges and unexpected events with confidence and security. Prudent planning, budgeting, and risk management strategies must be employed to prepare individuals and families against economic downturns, job loss, unexpected expenses, or expenses they did not anticipate. Building emergency savings may help. Other ways can include diversifying income sources, paying down debt faster or investing in long-term assets with long-term value. Staying aware of economic trends while seeking professional advice as needed and adapting financial plans as required can also assist individuals and families navigate times of financial uncertainty with confidence and stability.

## Practice of Permaculture and Sustainable Living

Sustainable living principles such as permaculture foster environmental stewardship, resilience, and self-sufficiency. Permaculture emphasizes working with nature to design systems that satisfy human needs while improving ecosystem health - practices like organic gardening, water conservation, renewable energy production and waste reduction are examples. By adopting permaculture principles individuals can cultivate resilient food systems while decreasing ecological footprint and strengthening community resilience - adopting sustainable living practices not only benefits the planet but can enhance personal well-being as well as increase resistance against challenges caused by environmental disruptions or disruptions.

## Enhancing DIY and Repair Skills

**Honing Do-It-Yourself (DIY) and Repair Skills is an empowering means of taking control over household** maintenance, saving both money and self-reliance. DIY skills may include carpentry, plumbing, electrical work, painting or general home maintenance tasks that individuals can learn through hands-on practice or online tutorials or community workshops - these allow individuals to overcome common problems quickly while engaging in home improvement projects that extend the lifespan of household items while simultaneously creating a sense of achievement, resourcefulness and resilience; equipping individuals to effectively respond when faced with unexpected emergencies!

## Implement Home Security Measures

Implementing home security measures is critical to protecting property, possessions and personal safety against intrusions or threats. This may involve installing security systems like alarms, surveillance cameras and motion sensors in the home to deter burglars and alert owners early of suspicious activity. Adding barriers such as sturdy locks on entry points reduces vulnerability for break-ins while adopting security conscious habits like keeping valuables out of sight, maintaining outdoor lighting systems, or participating in neighborhood watch programs can further deter criminal behavior while encouraging community security.

## Develop and Stockpile an Emergence Plan

Building and updating a bug-out plan and kit allows individuals and their families to safely evacuate during emergencies or disasters, improving chances of survival and recovery. A bug-out plan details evacuation routes, designated meeting points, essential items needed in specific risks/ scenarios as well as how and when they'll arrive. Emergency kits or "go bags," on the other hand, should contain necessities like food, water, clothing first aid supplies medications important documents tools etc. that should also be updated with changing circumstances as proactively preparing can reduce panic, confusion and risk during emergencies thus improving chances of survival and recovery.

## Understanding Psychological Responses to Crisis

Understanding psychological responses to crisis situations is integral for effectively supporting individuals and communities during difficult times. Crisis situations may prompt various emotional and behavioral responses such as fear, anxiety, anger and grief in individuals and communities alike. Recognizing that these responses are normal reactions to abnormal events vary significantly among individuals and should be provided psychoeducation and open dialogue, communities can reduce stigma while fostering resilience and increase resilience. An additional means of providing practical support, including counseling services, peer support groups and mental health resources helps individuals cope with stress and trauma more effectively. Forming strong social ties within communities further promotes psychological well-being and recovery efforts.

## Integrating Technology for Monitoring and Communication

Integrating technology for monitoring and communication enhances situational awareness while supporting effective emergency communications. Technologies such as sensors, drones, satellite imagery and Geographic Information Systems (GIS) enable real-time tracking of environmental conditions, infrastructure integrity and population movements. Communication tools such as social media, mobile apps, and emergency notification systems enable rapid dispersion of crucial information to affect populations. Emergency responders can utilize technology to enhance response coordination, allocate resources more effectively and increase public safety. But in doing so, it is imperative to address digital equity and privacy to ensure technology serves all members of society while upholding individual rights.

## Conduct Regular Drills and Training Exercise

Regular drills and training exercises are key for maintaining readiness and building capacity to respond appropriately in case of emergencies. These exercises simulate realistic scenarios, providing stakeholders the chance to practice response procedures, evaluate communication systems and pinpoint areas for improvement. By engaging diverse parties - emergency responders, government agencies, businesses, and community members alike - in drills, they foster greater cooperation and coordination between all involved. Debriefing sessions following exercises allow participants to evaluate their performances, share lessons learned, and refine emergency plans and procedures. Regular training drills bolster preparation, confidence and resilience ensuring swift responses when crises strike.

## Adaptation and Thriving in Shifting Environments

To adapt and thrive in ever-evolving environments requires flexibility, innovation, and resilience. As global trends shift and climate conditions alter, communities must anticipate and adjust accordingly in response to new threats or opportunities that emerge. Implementing sustainable practices, diversifying economic activities, and strengthening social networks. Furthermore, adopting innovation and technology may also assist adaptation by increasing resource efficiency, improving communication channels, and encouraging collaboration. Fostering an environment for learning, experimentation and continuous improvement helps communities navigate uncertainty in dynamic environments with ease and thrive through change and proactive responses to emerging issues that emerge - which ultimately ensures sustainable futures for generations yet unborn.

## DIY Project: Emergency Go-Bag Assembly

**Materials:**

- Backpack or durable bag.
- Water bottles or hydration pack
- Non-perishable food items (such as energy bars, canned goods)
- Multi-tool or Swiss army knife
- Flashlight with extra batteries
- First aid kit
- Whistle
- Emergency blanket or sleeping bag.
- Map of local area
- Personal hygiene items (toothbrush, toothpaste, hand sanitizer)
- Waterproof matches or lighter
- Cash (small bills and coins)
- Portable phone charger with cables
- Important documents (ID, passport, insurance information)
- Extra clothing (socks, underwear, jacket)
- Duct tape.
- Plastic bags (for waste disposal or waterproofing)
- **Notepad and pen**
- Emergency contact list
- Personal medications (if applicable)

**Procedure:**

- Preparing for the unexpected means creating an emergency go-bag - essential no matter the circumstance. Start by selecting a durable backpack or bag capable of accommodating essential items; ensure it can easily be reached in case of an unexpected crisis.
- Start by packing water bottles or hydration packs so you'll always have access to water during an emergency, while non-perishable food such as energy bars and canned goods provide sustenance without the need for cooking.
- Include in your first aid kit an indispensable multi-tool or Swiss army knife capable of performing various functions and an extra battery flashlight to provide illumination during power outages. A comprehensive first aid kit should also be present as this provides care for minor injuries that might arise.
- Whistles can be essential tools in signaling for help while an emergency blanket or sleeping bag provides warmth in cold conditions. Carry an accurate map of the local area with you so that unfamiliar terrain doesn't present itself unexpectedly.
- Personal hygiene items, such as toothbrushes, toothpaste, hand sanitizer and waterproof matches or lighters help ensure cleanliness while also helping stop germ spread. Waterproof matches or lighters make fire starting possible in case warmth or cooking are needed for warmth or cooking purposes.

- Cash in small bills and coins can come in handy during power outages or when electronic payment methods become inaccessible, while having access to a portable phone charger with cables ensures all communications devices remain active.
- Important documents such as identification cards, passports and insurance info should be securely kept within a go-bag for safety and convenience. Incorporating clothing, duct tape, plastic bags and notepad with pen will enhance comfort and functionality further.
- Make sure that the go-bag contains an emergency contact list as well as any medications necessary in an evacuation or emergency.
- By creating an emergency go-bag using these materials and following this procedure, you'll be better equipped for whatever challenges may come your way and ready for whatever may arise.

# BOOK 3: PREPAREDNESS HANDBOOK: CRISIS READY

Effective crisis management begins by identifying and prioritizing potential crises that might threaten an organization, through conducting thorough risk analyses to detect various threats and vulnerabilities from natural disasters to technological failures or geopolitical instability. Once identified, potential crises must then be prioritized based on likelihood of occurrence and their potential effect on key business functions, reputation, or stakeholders - this allows organizations to allocate resources strategically while developing targeted response plans designed to minimize disruption while mitigating risks and minimize disruptions.

## Customized Crisis Response Plans

Creating personalized crisis response plans is key for successfully handling and mitigating the effects of emergencies or disasters, including emergencies arising due to natural catastrophe. Such plans specify actions and procedures which must be carried out prior to, during, and post crisis situations to provide coordinated and successful responses to crisis events. Each organization requires its own customized crisis response plan tailored specifically for them based on factors like regulatory framework, geographic location, and stakeholder needs. Key components of an effective crisis response plan include clear roles and responsibilities, communication protocols, escalation processes and resource allocation strategies. Regular testing, training, and reviewing ensure readiness and efficiency when handling crises.

## Building Resilient Supply Chains

Building resilient supply chains is essential to business continuity and mitigating disruption caused by crises. This requires diversifying suppliers, creating alternative sourcing options, and implementing robust risk management practices to identify vulnerabilities, with technologies like blockchain providing transparency, traceability, and agility within supply chain operations. With resilient supply chains in place, organizations can anticipate disruptions faster, reduce downtime significantly more quickly while upholding customer satisfaction and loyalty more reliably than before.

## Implementation of Communication Strategies

Effective communication is paramount to managing crises and maintaining trust between stakeholders. Organizations should devise and implement tailored communications plans tailored specifically for various phases of crisis response - preparedness, response, and recovery. Establish clear channels of communication, identify roles and responsibilities for spokespersons, and deliver timely, accurate information to both internal and external stakeholders. Communication strategies must also address potential hurdles such as misinformation, rumors and media scrutiny while offering channels of feedback and dialogue for feedback and engagement. By adopting successful communication plans in crisis management efforts, organizations can enhance transparency, credibility, and resilience to avoid disaster.

## Establish Emergency Contact Protocols

Establishing emergency contact protocols is vital to timely and effective crisis communication efforts, including creating up-to-date lists of contacts for key personnel, stakeholders, external partners such as emergency services, regulatory agencies, and media outlets. Protocols should detail primary and alternate modes of contact such as phones calls/email/text messaging/social media to maximize redundancy and flexibility during crises situations. Regular testing helps identify any weaknesses and provide assurance of preparedness during an incident response situation.

## Drills and Exercise for Crisis Regular Conducting

Crisis drills and exercises should form the cornerstone of every organization's preparedness strategy. By simulating potential crises in controlled conditions, drills allow teams to practice responding more quickly under stress. Regular drills allow organizations to identify weaknesses in their crisis management plans and procedures, helping to refine and strengthen them over time. These drills also familiarize employees with their roles and responsibilities during an emergency, increasing coordination and efficiency when faced with real emergencies. Furthermore, drills provide organizations with an opportunity for feedback and evaluation so they can assess their readiness and make any necessary changes or modifications based on these evaluations.

## Emergency Employee Response Training

Employee Response Crisis Training for Management Teams is vital to guarantee rapid and efficient responses in emergencies. This specialized course equips managers with the knowledge, skills, and abilities required to lead their teams effectively during times of turmoil, such as communication protocols, decision-making processes, resource allocation strategies etc. By giving this type of comprehensive training to management teams during emergencies they will be empowered with making timely and informed decisions under pressure, minimizing disruptions while mitigating impacts caused by crises on businesses and its stakeholders.

## Maintain Continuity Business Plans

Maintaining continuity business plans is integral for building resilience and sustainability in times of crises, providing protocols and procedures for maintaining essential functions and operations during disruptions such as natural disasters, cyber-attacks, or pandemics. By regularly revising and updating plans, organizations can adapt quickly to ever-evolving threats and shifting business environments. This includes identifying key resources, creating alternate work arrangements, and taking precautionary steps to minimize disruptions while upholding service levels. Maintaining continuity business plans involves training employees, conducting drills, and cultivating an organizational culture of preparedness. Businesses who prioritize continuity planning will increase their capacity to withstand crises while emerging stronger afterward.

## DIY Project: Emergency Communication Plan Creation

**Materials:**

- Pen and paper or computer for documentation
- Contact information for family members, friends, and emergency services.
- Local maps
- Battery-operated or hand-crank radio
- Cell phone with chargers and backup batteries
- Paper and envelopes for written communication
- Internet access (if available)
- Emergency contact cards for each family member
- List of designated meeting places

- Whistle or signaling device.
- Personal identification documents (ID, passports)
- Important medical information and records
- Portable phone charger
- Waterproof container for storing documents and communication devices.
- Flashlight with extra batteries
- Notepad and pen for taking notes.
- Social media accounts for emergency updates
- Contact information for neighbors and community organizations.
- Emergency notification apps or services
- Two-way radios or walkie-talkies for short-range communication

**Procedure:**

- Communication during times of emergency is essential, which makes having a solid emergency communication plan even more crucial. Start by gathering all required supplies such as pens and papers or computers for recording purposes.
- Consolidate contact details of family, friends, and emergency services; make sure it's current and **easily accessible. Consider including local maps with evacuation routes or meeting places marked**.
- Make sure to have both a battery-operated or hand-crank radio available so you can receive emergency broadcasts and information. Also ensure your mobile phone battery remains fully charged; remember to include backup battery chargers.
- Prepare paper and envelopes in case electronic methods of communication become inaccessible; if internet access is available, take full advantage of it to stay informed and send updates to family and friends.
- Create emergency contact cards for each of the family members with pertinent contact info and designate meeting locations both inside and outside your neighborhood.
- Include in your kit a whistle or signaling device for alerting emergency services, when necessary, personal identification documents and medical records as well as emergency contact info and forms for easy transport.
- An emergency phone charger will keep your cell phone functional during extended power outages, and keeping documents and communication devices stored safely away in a waterproof container will protect them from water damage.
- Keep a flashlight with extra batteries nearby in case of nighttime emergencies, along with a notepad and pen to record important information and take notes as quickly as possible.
- Use social media accounts for emergency updates and communications among friends and family. Exchange contact info with neighboring community organizations to gain additional support in times of distress.
- Emergency notification apps or services offer real-time alerts and updates; two-way radios/walkie-talkies may provide short range communication among family or neighbors.
- By following these steps and creating an emergency communication plan with this material, you'll be better equipped to stay in contact and informed during any crisis.

# BOOK 4: CRISIS MIND: NAVIGATING MENTAL HEALTH

Understanding mental health during crises is vital to providing appropriate assistance to individuals facing difficulties. Crisis situations such as natural disasters or public health emergencies often bring elevated levels of stress, anxiety, depression, or trauma that require early intervention and supportive measures to restore well-being quickly and appropriately. Acknowledging the complexity of mental health in crisis contexts enables communities and organizations to tailor interventions according to each affected individual and foster resilience among affected persons.

## Recognizing Signs of Psychological Distress

Acknowledging signs of psychological distress is crucial for early intervention and support. Signs may manifest differently depending on an individual's unique circumstances or crisis but typically include changes to mood, behavior, sleep patterns and cognitive functioning as well as withdrawal from social interactions, difficulty concentrating or increased irritability. By being alert to these signals' friends, family, professionals can offer timely help as well as connect individuals to resources to meet their mental health needs more quickly and effectively.

## Coping Strategies to Manage Stress and Anxiety

Coping strategies to address stress and anxiety play an essential part in maintaining mental wellbeing during crisis situations, whether through mindfulness techniques, relaxation exercises, physical activity or engaging in enjoyable activities. People also benefit from seeking social support, living an overall healthier lifestyle, or practicing positive thinking as key coping mechanisms; by adopting such mechanisms they are better equipped to navigate crises as they come along and increase resilience against hardship.

## Seek Help: Resources and Hotlines

Individuals experiencing mental health challenges during a crisis should access support resources and hotlines as soon as they arise, including crisis hotlines, mental health helplines, counseling services and support groups. Trained professionals are on hand to offer immediate help while offering emotional support as they guide individuals toward appropriate interventions or treatments options. Accessible yet confidential support services provide individuals with all the help necessary for them to overcome mental health concerns while building resilience against future difficulties.

## Building Resilience with Positive Psychology

Building resilience through positive psychology involves honing strengths and developing an optimistic attitude to manage difficult circumstances effectively. It works by emphasizing personal development while emphasizing strengths over weaknesses - an approach which empowers individuals to overcome challenges, recover quickly from setbacks, and persevere despite hardship.

## Self-Care Techniques to Achie Mental Well-being

Self-care practices for mental well-being are essential in maintaining balance and strengthening resilience during times of turmoil, such as setting boundaries, prioritizing self-care activities, practicing mindfulness or relaxation techniques and nurturing supportive relationships. People also stand to benefit from leading healthier lifestyles - including regular physical activity, nutritious eating habits and adequate sleep - which all make their contribution towards managing stress more effectively. By including self-care into everyday routines individuals can replenish physical, emotional, and mental resources necessary for effectively dealing with uncertainty and stress more effectively.

## Supporting Others in Crisis

Assisting loved ones during times of distress requires empathy, patience, and effective communication. Listen without judgment as your loved ones share their fears or worries and validate feelings; provide practical assistance such as task help or transportation needs; encourage expression of emotions/thoughts by offering open channels; explore resources/support networks available and respect autonomy while gently leading towards effective coping mechanisms and professional help if required.

## Attaining Independence by Overcoming Stigma

Overcoming stigma surrounding mental health is essential in encouraging individuals to seek professional assistance during crises. By normalizing discussions about it and emphasizing its strength rather than weakness, encouraging open discussions on emotions and mental well-being while challenging misconceptions surrounding therapy or treatments and offering to accompany loved ones if they're anxious, individuals can feel more empowered to prioritize their mental wellbeing by seeking assistance they require. By decreasing stigma and increasing acceptance, individuals will feel armed to prioritize their mental well-being by seeking assistance they require.

## Create a Safe and Supportive Environment

Establishing a safe and supportive environment is vital in aiding healing during crises. Make sure loved ones feel physically and emotionally safe by maintaining confidentiality, respecting boundaries, avoiding judgment or criticism and prioritizing open dialogue and trust between relationships, prioritizing active listening and empathy, encouraging self-care practices such as providing meals, transportation or childcare and offering practical help such as transportation services if required; additionally create opportunities for relaxation like taking leisure activities together or spending quality time outdoors - these nurturing environments give loved ones strength to navigate challenges more successfully

## Mindfulness and Meditation Techniques

Mindfulness and meditation techniques can be highly effective tools for relieving stress during times of crisis. Encourage loved ones to incorporate these strategies by practicing being present without judgment in each moment as well as various meditation techniques like deep breathing exercises or guided imagery into daily activities like eating mindfully or practicing gratitude - to build resilience against crisis situations more efficiently.

## Manage Trauma and Post-Traumatic Stress Disorder (PTSD)

Facilitating recovery from trauma and posttraumatic stress disorder (PTSD) requires professional assistance from mental health providers who specialize in trauma treatments. Encourage loved ones to reach out for professional assistance from trauma treatment experts. Support them emotionally while becoming educated on symptoms and triggers associated with posttraumatic stress disorder (PTSD). Respect their boundaries and triggers, while refrain from pressuring them into discussing traumatic experiences prematurely. Encourage healthy coping mechanisms such as relaxation techniques, support groups and trauma-focused therapies (CBT or EMDR are among several). By offering understanding and support to loved ones living with trauma and posttraumatic stress disorder (PTSD), you can help them navigate its complex terrain in search of healing.

## Addressing Substance Abuse and Addiction Issues

Combatting substance abuse during an emergency requires taking an integrated approach that includes medical treatments, counseling services and support services. Encourage those struggling with substance use disorders to seek professional assistance through addiction specialists or treatment centers. Avoid providing judgmental support or enabling behaviors which might feed into addiction and seek support groups, counseling services and rehabilitation programs as resources to create personalized recovery plans for loved ones with substance use issues. Also prioritize self-care while caring for someone experiencing substance use problems as it can be emotionally straining.

## Balancing Work and Personal Life during Crisis

Balance work and personal life during times of crisis requires flexibility, time management, and self-care. Set priorities according to priority and urgency while communicating openly about any challenges or limitations you are encountering from employers or colleagues. Set boundaries between work and personal lives to prevent burnout while maintaining well-being - schedule regular breaks, participate in enjoyable and relaxing activities such as sports and hobbies while seeking support from friends, family, or mental health professionals **as necessary if required - remembering it's okay to reach out for assistance or adjust expectations** during such times of chaos!

## Understanding the Impact of Crisis on Children and Adolescents

Understanding how crises affect children and adolescents is paramount to providing appropriate support and intervention measures. Young people may experience fear, anxiety, confusion, or sadness during times of distress. Create an atmosphere in which they feel safe to express their emotions and ask questions without feeling judged, provide age-appropriate information on the crisis and assure them of their wellbeing and safety. Monitor for signs of distress, such as changes in behavior or sleep patterns, and seek professional assistance if necessary. Furthermore, encourage healthy coping strategies like using art and play therapy as expressive outlets; maintaining routines; staying connected to friends and family.

## Cultivating Healthy Relationships and Boundaries

Cultivating healthy relationships and boundaries are crucial components of maintaining wellbeing during crisis times. Prioritize open and honest communications with loved ones and set clear boundaries to protect both physical and emotional well-being. Practice active listening, empathy, and respecting other's boundaries as you develop supportive relationships which offer encouragement, validation and understanding - distancing yourself from toxic or unhealthy dynamics while setting aside time for meaningful connections with loved ones while prioritizing activities which strengthen them further by cultivating mutual trust and respect between you.

## Exploring Alternative Therapies and Healing Modalities

Alternative healing modalities and therapies may provide valuable supplementary methods of wellness during times of distress or crisis, providing holistic recovery. Try including practices like acupuncture, massage therapy, mindfulness meditation or aromatherapy into your self-care routine to foster holistic recovery and promote holistic wellbeing. Explore creative therapies such as art therapy, music therapy or dance therapy as a means of releasing feelings and relieving stress. Additionally, nature-based therapies like ecotherapy or horticulture therapy may promote relaxation while deepening connection to nature - contributing towards overall well-being and relaxation. It's always wise to consult qualified practitioners or professionals when considering alternative therapies as alternatives may not always meet individual requirements and circumstances.

## DIY Project: Mind Matters: Navigating Mental Health in Crisis

**Materials:**

- Journal or notebook.
- Pen or pencil
- Relaxation and mindfulness resources (e.g., meditation apps, calming music)
- Emergency contact list
- List of local mental health resources (e.g., hotlines, support groups)
- Stress-relief toys or tools (e.g., stress balls, fidget spinners)
- Comforting items (e.g., a favorite book, cozy blanket)
- Art supplies (e.g., coloring books, colored pencils)
- Physical activity resources (e.g., yoga mat, workout videos)
- Healthy snacks and drinks
- Scented candles or essential oils for relaxation
- Self-help books or articles on mental health and crisis management
- Educational materials on coping strategies and resilience-building techniques
- Communication tools for staying connected with loved ones (e.g., phone, internet)
- Emergency medication (if applicable)
- Safety plan for managing crisis situations.
- List of affirmations or positive reminders
- Sleep aids (e.g., eye mask, earplugs)
- Guided imagery recordings or relaxation scripts
- Professional therapy or counseling resources (e.g., online therapy platforms, therapist contacts)

**Procedure:**

- Navigating mental health issues during times of distress requires meticulous self-care and coping mechanisms. Start by designing a space where your mental health toolkit will always be accessible in times of stress.
- Use your journal or notebook as an outlet to record thoughts, feelings and experiences that matter to you. Writing can be an excellent way to process emotions while documenting mental health journey.
- Implement relaxation and mindfulness into your everyday routine by making use of resources like meditation apps or soothing music to center yourself and reduce stress levels.
- Prepare an emergency contact list consisting of reliable friends, family, or mental health professionals that you can turn to during difficult times. Also gather details of local mental health resources like

hotlines or support groups which might offer additional help - they are key resources that will provide additional relief during challenging moments.
- Keep stress-relief toys or tools handy to alleviate anxiety and tension, such as your favorite book or cozy blanket, solace you during times of distress and provide much-needed solace and solace.
- Doing creative activities such as drawing or coloring to release emotions and foster self-expression can help. Physical exercise also aids mental wellbeing; incorporate yoga mats or workout videos into your schedule for maximum mental wellbeing.
- Maintain a supply of healthy snacks and drinks to provide nourishment to both body and mind during times of increased stress. Scented candles or essential oils may create an ambience conducive to relaxation and peace, creating the right atmosphere to achieve this end.
- Acquaint yourself with coping and resilience-building techniques through self-help books, articles, or online sources. Create an emergency safety plan for managing crises effectively; know when seeking professional assistance is appropriate and when seeking additional support is warranted.
- Utilize positive affirmations and reminders to foster resilience and hope, while including sleep aids or relaxation techniques into your bedtime ritual for restful slumber and rejuvenation.
- Use guided imagery recordings or relaxation scripts to ease anxiety during moments of high stress, while staying connected with loved ones using communication channels like phone calls and video chats - **creating a strong network and sense of support and affection from around the globe.**
- **Prioritize your mental wellbeing by seeking professional therapy or counseling, as necessary.** Remember it's okay to ask for help and taking good care of yourself is integral for navigating crises with resilience and strength.

# BOOK 5: WATER MASTERY: SURVIVAL TECHNIQUES

Hydration is of paramount importance in survival situations as it helps sustain bodily functions and overall health. Dehydration can quickly lead to serious consequences including fatigue, confusion, heatstroke or even death; sweat, respiration or other means can rapidly drain away water supply through sweat glands causing rapid dehydration; therefore replacing lost fluids is imperative to avoid dehydration and sustain physical and cognitive performance in extreme situations like outdoor adventures or emergencies where we lose water through sweat, respiration or other means - replenishment is imperative to sustain physical and cognitive performance! Additionally proper circulation temperature regulation as well as waste removal are all vitally essential factors to survival when confronted by such hostile environments!

## Recognizing Safe Water Sources in the Wild

Locating safe water sources in the wild is crucial to avoid illness or disease, and should include natural sources like streams, rivers, lakes, and springs as they tend to contain clean water. Although they might seem safe at first glance, even these seemingly idyllic sources could harbor bacteria, parasites or chemical pollutants which pose risks - therefore it's necessary to evaluate water quality based on proximity factors like human activity levels, animal presence levels and visible contamination indicators; when in doubt prioritize sources that appear clear with no visible debris present when in doubt.

## Methods for collecting and storing water.

Survival situations call for reliable water collection and storage systems that ensure access to clean drinking water sources, using containers such as canteens, water bottles or collapsible water bladders for collection from natural sources - using containers made from leaves, bark or animal hide may even work! To maximize long-term freshwater resources, consider creating rainwater collection systems using waterproof materials like tarpaulins ponchos to collect precipitation as rain. Once collected water has been safely secured to prevent contamination or loss.

## Techniques for Water Filtration and Purification

Filtration and purification techniques are crucial in making water safe for consumption by eliminating contaminants that pose health threats. Filtration methods use porous media such as cloth filters or ceramic filters to rid water of particles such as dust or sediment particles as well as bacteria or microorganisms that might pollute it. Purification methods like boiling, chemical treatment or ultraviolet (UV) sterilization aim to identify and eliminate or inactivate harmful pathogens like bacteria, viruses and parasites that threaten public health. Combination methods, like employing both portable water filters and chemical treatment processes to purify drinking water safely are recommended as additional safeguards against contaminants or pathogens present. Prioritize using reliable purification mechanisms so your drinking water remains free from contaminants or pathogens.

## Building DIY Water Filtration Systems

Water filtration can be an efficient and economical means of providing clean drinking water in emergency or outdoor settings, using readily available materials like gravel, sand and activated charcoal layers in containers like plastic bottles or buckets as filters to remove impurities and improve water quality. Although DIY systems may not offer as effective purification solutions as commercial filters do when other options are unavailable - yet another useful way of purifying drinking water.

## Utilizing Natural Water Purification Techniques

Natural water purification techniques harness environmental processes to increase both quality and safety of drinking water supplies. Exposing it to ultraviolet (UV) rays from sunlight may assist with breaking down certain contaminants through exposure, while vegetation and soil act as natural filters, trapping sediment and impurities as it passes through. Furthermore, placing sources upstream from human activity or animal habitation reduces contamination risks further; although natural purification methods might not completely remove all potential pollutants altogether, they serve as complementary measures that help enhance emergency situations when there's no access to clean drinking water sources.

## Emergency water treatment chemicals and tablets

Emergency water treatment chemicals and tablets provide an ideal portable solution for purifying water during survival situations. Typical products contain disinfectants like chlorine dioxide or iodine which effectively kill bacteria, viruses, and protozoa in untreated water sources. Simply add the appropriate dose to untreated sources before waiting the prescribed contact time before drinking or bathing - while chemical treatments work at killing pathogens, they may leave an unpleasant flavor or smell behind - those allergic or **suffering thyroid conditions should seek alternative approaches as treatment solutions.**

Boiling Water safely requires considerations that go beyond its technical components. Below are best practices and considerations related to boiling water at home or for work purposes.

Boiling water can be one of the most reliable emergency methods of purifying it, killing most pathogens such as bacteria, viruses, and parasites. To boil, bring to a rolling boil for at least one minute (three at higher altitudes) before allowing to cool off before cooling it further. While boiling is highly effective at killing pathogens it does not always remove certain chemical contaminants or toxins; additionally boiling requires both fuel source and source for optimal performance - however when conducted appropriately it remains an easily accessible means for guaranteeing water safety.

## Solar Water Disinfection System

Solar Water Disinfection (SODIS) uses sunlight to disinfect water and make it safe for consumption. To use SODIS, fill a clear container such as plastic bottle with untreated water and place it directly under direct sunlight for at least six hours (or longer if clouds cover part of sky), so UV radiation from sun kills bacteria, viruses and parasites present in it - rendering it safe to consume and drink! Although SODIS offers a simple and environmentally friendly treatment approach, its efficacy may depend on weather conditions as well as water clarity - thus it may not work effectively on all contaminated waters due to this method's limited effectiveness or efficiency depending on conditions that influences weather and clarity factors.

## Distillation: Purifying Water Through Evaporation

Distillation is an efficient method for purifying water by extracting contaminants through evaporation and condensation. Distillation works by heating the liquid to its boiling point, which then causes impurities such as salts, heavy metals, microorganisms to evaporate out as steam before condensing back to liquid form creating clean drinking water for consumption. Distillation removes even heavy contaminants making it suitable for purifying heavily polluted or saline sources with ease; although energy to heat water must first be applied it provides safe drinking water that should ensure safe drinking water when needed both emergencies as well as everyday consumption.

## Understanding Waterborne diseases and prevention

Understanding waterborne diseases and prevention strategies are fundamental for protecting public health and well-being. Waterborne illnesses are caused by pathogens such as bacteria, viruses and parasites infiltrating water sources contaminated by pathogens like bacteria or virus and leading to illness or infection when consumed; common examples include cholera, typhoid fever, giardiasis and cryptosporidiosis - common ones can include accessing safe drinking water with proper hygiene practices employed; also implemented water treatment and sanitation systems as prevention measures against waterborne illnesses as education on risks relating to safe water practices is key in terms of decreasing incidence while improving community wellbeing.

## Calculating Water requirements for survival

Calculating water needs for survival is vitally important to ensuring adequate hydration during emergency situations or outdoor activities, avoiding dehydration. Factors which influence our water needs include environmental conditions, physical exertion levels, age, body size and health status of everyone. As a general guideline aim for at least 2 liters (64 ounces) daily consumption - although in hot or arid climates or during strenuous physical activities this could increase significantly; thirst cues, urine color changes and overall wellbeing - always prioritize adequate hydration over dehydration related risks!

## Hydration Strategies in Extreme Environments

Hydration strategies in extreme environments must be developed carefully with consideration given to environmental conditions and climate. When dealing with hot or arid climates, prioritize water conservation measures by limiting excessive sweating and physical exertion during peak heat periods of each day. Be sure to drink small, frequent sips of water throughout the day to replenish electrolytes lost through sweat, as this will keep hydration levels optimal and avoid dehydration. Consider supplementing this water consumption with electrolyte-rich beverages or oral rehydration solutions which contain valuable minerals which will replace lost minerals while helping prevent dehydration. Dehydration in cold environments is common due to increased respiratory water loss and decreased thirst sensation; so, drinking regularly, even when not thirsty is crucial. Monitor hydration status closely and adjust fluid intake based on individual circumstances for maximum performance in extreme environments.

## DIY Project: Water Worries: Mastering Survival Hydration Techniques

**Materials:**

- Water purification tablets or drops.
- Portable water filter or purifier
- Collapsible water container or canteen
- Stainless steel water bottle
- Distillation kit (e.g., solar still)
- Plastic tubing or hose

- Bucket or large container.
- Tarp or plastic sheeting
- Collection vessel (e.g., empty bottles, jars)
- Handheld shovel or digging tool.
- Waterproof tape
- Solar water disinfection (SODIS) bags or containers
- Coffee filters or cloth for pre-filtering water
- Large cooking pot or kettle
- Sunlight (for SODIS method)
- Fire-starting materials (e.g., matches, lighter)
- Water-resistant storage containers
- Water collection guidebook or manual
- Rope or paracord (for constructing solar still or rainwater collection system)
- Emergency whistle or signaling device.

**Procedure:**

- Mastering survival hydration techniques is crucial to providing access to clean drinking water during crisis situations. **Start by exploring various water purification and collection methods to meet your hydration requirements.**
- Water purification tablets or drops can help eliminate pathogens found in untreated water supplies. Be sure to follow all instructions when applying these chemicals for optimal disinfection results.
- Portable water filters or purifiers offer an on-the-go solution for filtering out contaminants and bacteria from natural water sources, providing clean drinking water without contaminants and bacteria contaminating it. Find one that meets all your specific requirements to suit both your lifestyle needs and usage expectations.
- Carry a collapsible water container or canteen when transporting and storing purified water for safekeeping in survival situations. Stainless steel water bottles also make excellent choices to carry water safely and reliably.
- For distilling water, invest in a distillation kit such as a solar still. Setup this setup using plastic tubing or hose to collect evaporated water that then channels into an appropriate vessel for collection.
- In situations of limited water access, collecting rainwater through rainwater collection techniques such as setting up a tarp or plastic sheeting to catch rainfall can help. Once collected, use buckets or large containers to collect runoff, before storing it safely for future use.
- Solar water disinfection (SODIS) is another means of purifying water using sunlight. Fill SODIS bags or containers with untreated water and leave it under direct sunlight for several hours to kill off microorganisms and harmful organisms that are present.
- Pre-filtering water with coffee filters or cloth can remove sediment and debris before purification, and boiling in large kettle is another proven strategy for eliminating pathogens and making drinking water safe to drink.

- Under emergency conditions, knowing how to start a fire using matches or lighters is imperative in boiling water and purifying it through heat treatment.
- Store purified water in water-proof containers to reduce contamination and ensure long-term storage. Consult a water collection guidebook or manual for detailed instructions on various purification and collection techniques.
- Construction of a solar still or rainwater collection system often requires rope or paracord for securing materials and creating a sturdy structure, in addition to carrying an emergency whistle or signaling device for alerting rescuers in case of emergencies.

# BOOK 6: PANTRY PREP: FOOD SECURITY MASTERY

Reviewing your food supply is the first step toward food security and emergency preparedness. Conduct an inventory of perishable and nonperishable items you have on hand as well as emergency rations or long-term food storage supplies; factor in things such as expiration dates, nutritional values and any dietary preferences or restrictions you might need for emergency situations; evaluate adequacy according to household size, duration of needs expected as well as potential disruptions that might threaten access; pinpoint gaps or deficiencies to guide planning efforts while prioritizing necessary actions towards food security improvements.

## Understanding Food Security: Exploring Its Significance

Food security is vital to individual and community well-being and resilience. This term encompasses access to enough nutritious food that meets dietary preferences as well as being obtained and consumed sustainably and without harm to our environment. Food security helps promote physical and cognitive function as well as overall productivity - essential factors during times of uncertainty or crisis. By prioritizing it, people and communities can reduce hunger risks such as malnutrition as well as build long-term foundations of wellbeing that ensure long-term wellbeing and prosperity for themselves and future generations.

## Establish a Comprehensive Food Storage Strategy

**Creating an effective food storage plan is key to building resilience and preparedness during emergencies** or disruptions to food access. Start by assessing your household's dietary requirements, family size and anticipated duration of need to create your food storage strategy. Establish an inventory of essential food items, including staple grains, proteins, fruits, and vegetables as well as nonperishable foods; create target stockpiling quantities to cover both short-term and long-term needs; stockpile your emergency supply before it runs out! Before selecting and organizing food items and organizing their storage, take into consideration factors like space requirements, shelf life and rotation practices. Review your plan frequently to ensure its relevance and effectiveness in meeting household requirements.

## Stockpiling Long-term Food Staples

Stockpiling long-term food staples is an integral component of food storage plans, providing sustenance and nutrition during extended periods of food scarcity or disruption. Prioritize purchasing shelf-stable, long shelf life foods like grains (rice, pasta and oatmeal); legumes (beans and lentils); canned goods containing vegetables fruits proteins etc.; dried fruits nuts seeds as well as freeze dried or dehydrated products to build your stockpile with essential vitamins minerals macro nutrients to support health over time - choose those rich in essential vitamins minerals macro nutrients such as vitamins minerals macro nutrients or freeze dried/dehydrated items regularly rotate stocking it to maintain quality!

## Establish an Emergency Food Pantry

Creating an emergency food pantry is an efficient and practical solution to ensure access to nutritious food in case of short-term emergencies or crises. Find an appropriate storage location within your home - pantry, closet or basement are good examples - for this essential inventory management task and organize accordingly for easier access and inventory tracking. Stock your pantry with non-perishable foods that require minimal preparation or can be consumed without electricity or cooking facilities, such as canned goods, dried fruits, nuts, nut butters, whole grain crackers, shelf stable milks and meal replacement bars. Consider also including emergency cooking supplies like a portable stove, fuel, and cooking utensils to prepare meals in case they're needed in an emergency. Regularly assess and replenish your emergency food pantry to stay prepared in times of change and uncertainty.

## Rotating and Managing Food Inventory

Rotating and managing food inventory is critical to ensure freshness, quality, and safety in any food storage system. Use the "first in, first out" (FIFO) principle to make sure older items are used or consumed before newer ones arrive in storage; check expiration dates regularly as well as inspect packaging for signs of damage or spoilage; keep track of your inventory using an Excel spreadsheet or inventory log; assess consumption patterns regularly as needs shift or expirations dates near; adjust stockpile accordingly in response. By practicing meticulous inventory management, you can minimize food waste as well as ensure space utilization is optimized while your emergency supply stays full stocked for emergency use - saving valuable storage space as well as time allowing your stockpile remains well stocked and prepared when an emergency strikes.

## Proper Packaging and Storage Techniques for Food Products

Food packaging and storage techniques are critical in protecting the freshness, flavor, and nutritional value of stored items. When choosing packaging materials that protect from contamination by airtight seals that block airflow or moisture entry as well as pesticide residue, use food-grade containers (plastic bins, glass jars or Mylar bags are perfect) tightly sealed against air or moisture entry to store dry goods or bulk items; perishable items (grains legumes or dried fruit should be stored somewhere cool, dark and dry - this helps extend shelf life while maintaining quality over time while managing inventory management through rotational practices and inventory rotation techniques. Label containers accordingly for inventory control and rotation purposes and facilitate inventory management/rotation efforts!

## Utilize Freeze-Dried and Dehydrated Foods

Integrating freeze-dried and dehydrated food storage solutions offers many advantages, including extended shelf life, lightweight storage solutions, flavor retention and nutrition retention. Dehydration processes remove moisture from foods to inhibit bacterial growth and spoilage - they remain shelf stable over extended time periods before needing rehydrating with water before being eaten again when ready. Store frozen fruits, veggies, meats, dairy products, or prepared meals among your food storage options to expand variety and nutrition in emergency pantry storage situations.

## Canning and Preserving Fresh Produce

Canning fresh produce is an old and proven technique for prolonging its shelf life and decreasing food waste. Canning involves heating food in sealed jars to kill bacteria and form an airtight vacuum seal to avoid spoilage; for best results use top-grade produce harvested at peak ripeness for this method and adhere to approved canning recipes and techniques; alternatively try pickling, fermenting, drying or freezing to preserve its flavors and nutritional values long-term storage - properly preserved canned goods can provide invaluable sources of nutrition as well as flavor for long-term storage systems alike!

## Understanding Expiration Dates and Shelf Lives

Understanding expiration dates and shelf lives are integral for optimizing food storage practices, and to guaranteeing food safety and quality. Expiration dates represent when an item should remain safe to consume when stored properly; differentiating between different kinds of expiration dates such as "sell by," "use by," and "best by," can vary significantly in meaning and significance; generally trust your senses and judgment when judging quality and freshness when making assessment calls, prioritize proper storage rotation practices to prolong shelf life and minimize waste, discard any food that shows signs of spoilage such as unusual odor, appearance or texture to protect yourself against foodborne illness!

## Establish a Garden to Harvest Fresh Vegetables

Establishing a garden to source fresh produce can be both rewarding and sustainable; you will increase the variety and nutritional content of the foods in your diet by choosing an area with ample sunlight and well-draining soil for your plot. To start growing healthy fruits, vegetables, herbs - and more! - choose an appropriate location. Begin small with easy-to-grow crops like tomatoes, lettuce, peppers, and herbs and gradually expand as your skills and confidence increase. Integrate organic gardening methods like composting, mulching, and natural pest control methods into your gardening practices for improved soil health and biodiversity. Regularly water, weed and maintain your garden for optimal growth and productivity. By producing fresh produce yourself, you can enjoy tasty dishes while simultaneously decreasing environmental **footprint and strengthening connections to nature.**

## Acquisition of Hunting and Foraging Skills

Learning basic hunting and foraging skills can open numerous avenues to supplementing your food supply with game, fish, and edible plants from nature. Begin your hunting expedition by becoming familiar with local regulations and ethical hunting practices; seek guidance and training from veteran hunters or outdoor organizations. Learn to recognize common game species, tracks, and signs of wildlife activity as you practice essential hunting techniques such as stalking, tracking and field dressing. Education on local edible plants, mushrooms and foraging opportunities will allow you to avoid poisonous or harmful species when collecting wild foods from nature. By developing hunting and foraging skills you will increase your self-reliance and resilience when accessing sustenance from nature.

## Implementation of Permaculture Practices

Implementing permaculture practices can ensure sustained food production while simultaneously supporting ecological resilience and biodiversity. Permaculture is an approach which mimics natural ecosystems to design productive yet regenerative food systems. Implement principles like polyculture planting, companion planting and agroforestry to maximize yields while improving soil fertility and conserving water resources. Utilize mulching, composting and rainwater harvesting techniques to build healthy, resilient soils while decreasing dependence on external inputs. Employ diversity-enhancing practices within your garden or food production system to strengthen resilience against pests, diseases, and climate variability - by adopting permaculture principles you can develop abundant and resilient food systems which offer long-term sustenance to you and your community.

## Exploring Alternative Protein Sources

Diversifying your diet by exploring alternative protein sources is one way of diversifying it and decreasing reliance on traditional animal products while fulfilling nutritional needs. Consider including plant-based protein sources in your meals such as legumes (beans, lentils, and chickpeas), nuts and seeds, tempeh, tofu, or seitan as sources. Experiment with different cooking techniques and recipes until you develop delicious dishes tailored specifically to your palate! Explore alternative protein sources such as edible insects, algae and cultured meat that provide sustainable and environmentally friendly ways of producing protein. By diversifying your protein consumption sources you can enhance nutritional quality while simultaneously decreasing environmental impact and creating food security for future generations.

## Address Dietary Restrictions and Allergies in Emergency Planning

Emergency planning must include provisions to address dietary restrictions and allergies to safeguard individuals with special dietary needs and requirements. Locate and document any dietary restrictions or allergies within your household or community, such as food intolerances or allergies to specific ingredients as well as vegetarianism/veganism preferences. Prioritize emergency food supplies and meal plans which meet special dietary needs when creating emergency supplies and meal plans; for instance, allergen-free options, gluten-free grains, dairy alternatives, or plant-based protein sources should all be prioritized when creating emergency supplies and meal plans. Communicate any dietary restrictions clearly to emergency responders, shelters, or community organizations so that appropriate accommodations and support during crises are met.

## Make Emergency Meal Plans

Meal planning for emergencies helps ensure you will have enough food supplies and nutritionally balanced meals when times become difficult or disrupted. Create meal plans using shelf-stable, nonperishable items from your emergency food supply such as grains, legumes, canned goods, dried fruits nuts seeds. When designing meal plans for your household consider factors like dietary preferences cooking capabilities nutritional needs as well as simple versatile recipes which require minimum prep/cook time as well as prioritizing nutritious items to support health & well-being during times of emergency situations.

## Utilizing Community Resources for Food Security

Utilizing community resources for food security can offer tremendous help during times of hardship and need. Check your local food banks, pantries, soup kitchens and community gardens to gain emergency food aid or access fresh produce for meals; volunteer your time or resources in support of food distribution efforts to bolster community resilience; engage with organizations like churches school's neighborhood associations in emergency preparedness initiatives to share resources among members while creating networks of support around vulnerable individuals and families in your neighborhood.

## Bartering for Food in Crisis Situations

Bartering for food during an emergency may be an efficient and economical means of meeting basic needs when traditional sources are limited or unavailable. Locate items in your possession which could be useful as trade or exchange items, such as extra food, water, medical supplies, tools, or skills. Develop relationships and networks within your community or among fellow prepper groups to facilitate bartering or trading arrangements during emergencies. Apply principles of fairness, transparency and reciprocity when conducting transactions and prioritize meeting all party's needs. By employing mutual aid and cooperation principles in times of crisis you can strengthen community resilience and adaptability.

## Emergency Cooking Methods Without Electricity

Emergency cooking methods without electricity enable you to safely prepare hot meals during power outages or off-grid situations, with various techniques like:

Propane or butane camping stoves; charcoal/wood-fired grills (wood/charcoal hybrid); portable outdoor stoves or rocket stoves, solar cookers/ovens.

DIY solar-powered cookers crafted with reflective materials for outdoor cooking; alcohol stoves or chafing dishes for use indoors; Dutch ovens or cast-iron skillets suitable for outdoor cooking, thermal cookers that retain heat during slow cooking processes, thermal bags for thermal cooking bags insulated to retain the heat - these all can serve to provide solar cooking on demand!

Practice these emergency cooking methods before an actual crisis strikes and be sure to have sufficient fuel or resources on hand in the event you must continue operations for extended periods.

## Continued Update and Revamp of Food Security Plan

Maintaining and revising a food security plan over time is essential to adapt to changing circumstances, needs and resources. Conduct regular inspections and evaluations of your emergency food supply, meal plans, and inventory management systems to identify gaps, deficiencies, or areas for improvement. Staying aware of threats or disruptions affecting food security in your region and adapting your preparedness strategies accordingly is critical for food security in any community. Participation in ongoing education, training and skill-building activities regarding storage, preservation and self-reliance of foods is also beneficial for maintaining food security in any situation. Work collaboratively with household members, neighbors, and community partners in your local area to strengthen food security and resilience. Being proactive yet flexible when planning food security can boost readiness and adaptability in response to unexpected change and uncertainty.

## DIY Project: Food Security: Preparing Your Pantry for Any Eventuality

**Materials:**

- Shelf-stable canned goods (e.g., beans, vegetables, fruits, soups)
- Dried grains (e.g., rice, pasta, oats)
- Canned protein sources (e.g., tuna, chicken, beans)
- Shelf-stable dairy products (e.g., powdered milk, evaporated milk)
- Cooking oil (e.g., vegetable oil, olive oil)
- Dried fruits and nuts for snacking
- Shelf-stable condiments (e.g., salt, pepper, spices, sauces)
- Jarred or canned sauces (e.g., pasta sauce, salsa)
- Boxed or canned broth or stock
- Shelf-stable bread or crackers
- Instant coffee or tea

- Nutrient-rich supplements (e.g., vitamins, protein powder)
- Shelf-stable baby formula (if applicable)
- Ready-to-eat meals (e.g., canned chili, pre-packaged meals)
- Freeze-dried or dehydrated foods
- Pet food (if applicable)
- Water storage containers
- Manual can opener
- Vacuum sealer and bags for food preservation
- Emergency food ration bars or meals-ready-to-eat (MREs)

**Procedure:**

- Prepping your pantry to meet any emergency is critical to ensure food security during emergencies or crises. Start by conducting an inventory review of current supplies to identify gaps or shortages that might exist in them.
- Stock up on shelf-stable canned foods such as beans, vegetables, fruits, and soups - they provide essential nutrition for healthy meals!
- Add dry grains such as rice, pasta and oats as versatile staples that can be used in numerous recipes. Tuna, chicken, and beans canned protein sources make invaluable additions to your pantry.
- Don't overlook shelf-stable dairy products such as powdered and evaporated milk that can be used in cooking or as substitutes for fresh dairy products.
- Cooking oil is essential in food preparation and adds both flavor and body to meals. Keep plenty of vegetable or olive oil on hand for cooking and baking projects.
- As snacks, dried fruits and nuts offer healthy options that require no refrigeration - perfect for when time is short!
- Make sure that you stock a wide variety of shelf-stable condiments like salt, pepper, spices, and sauces to enhance the flavors in your meals. Jarred or canned sauces such as pasta sauce or salsa can add depth and complexity to dishes.
- Broth or stock from either boxed or canned sources is invaluable when creating soups, stews, and sauces, while shelf-stable bread or crackers make convenient accompaniments to soups or spreads.
- Don't overlook beverages - include instant coffee or tea for an energy boost and extra hydration!
- Assuring you meet all your nutritional requirements in times of scarcity is easier by stocking up on nutritional supplements like vitamins or protein powder in advance.
- If you have infants or small children, stockpiling shelf-stable baby formula will ensure their nutritional needs are fulfilled.
- Ready-to-eat meals like canned chili or other prepackaged items provide fast, easy meals in times of emergency.
- For long-term storage needs, consider freeze-dried or dehydrated food items as these have long shelf lives and maintain nutritional benefits.
- For your pets' wellbeing in emergencies, do not forget to stock a supply of pet food.

- Store water in designated water storage containers to ensure an ample supply for drinking and cooking needs, using manual can openers in lieu of electricity to open cans when opening canned products.
- Consider investing in vacuum sealers and bags as part of your food preservation strategy to extend shelf life and prevent spoilage of perishable items.
- Finally, emergency food ration bars or meals-ready-to-eat (MREs) should be considered a last resort option during extended emergencies for providing sustenance.
- By taking these steps and stocking your pantry with these materials, you will ensure food security for yourself and your loved ones in any circumstance.

# BOOK 7: PREPPER'S RECIPE COLLECTION: SURVIVAL CUISINE

Staple Recipes from Pantry Supplies Prep food that will always have something on the table by using staple pantry supplies when creating staple recipes from pantry supplies - especially during times of uncertainty! Here are a few versatile dishes made using long-lived pantry items.

## Vegetable Soup

**Ingredients:**

Canned tomatoes, canned beans, dry lentils, pasta/rice, onion, and garlic with assorted dry herbs/spices added, as necessary.

**Directions:**

Start by sautéing onions and garlic before adding canned tomatoes, broth (or water), canned beans, dried lentils, and any vegetables available - season to your own liking with herbs and spices and simmer until everything is tender!

## Pasta Primavera

**Ingredients:**

canned chickpeas, canned tomatoes, coconut milk, onion, and garlic sauteing as you add curry spices such as turmeric cumin coriander etc. Simmer until all flavors blend before adding chickpeas for simmering until flavors develop further. This dish makes an impressive meal.

**Directions:**

Preheat pasta according to package instructions before sautéing onions and garlic until translucent; stir in canned tomatoes/sauce/veggies before tossing together with cooked pasta and garnishing with herbs as desired.

Rice and Beans for this Recipe consists of rice, canned beans, onion/garlic mixture, cumin (chili powder/paprika) when cooking rice and sautéing onions and garlic before adding canned beans with spices to sauté before topping with rice for serving.

## Tuna Casserole

**Ingredients:**

Make up this tasty casserole: pasta, canned tuna in water with cream of mushroom soup added, canned peas from can and breadcrumbs as a top layer.

**Directions:**

Cook pasta according to package instructions; combine with canned tuna, soup, and peas before topping with breadcrumbs and baking until bubbly.

Preheat quinoa according to package instructions; mix cooked quinoa with drained canned beans, corn, diced tomatoes and chopped vegetables before tossing with vinaigrette dressing. Black Bean Quesadillas

## Black Bean Quesadillas

**Ingredients:**

Contain Canned black beans as well as tortillas cheese onion garlic spices such as cumin chili powder.

**Directions:**

Mash canned black beans with spices before layering onto tortillas with cheese, onions, and garlic before cooking in a skillet until cheese melts and tortillas become crispy. Our recipes can easily adapt according to what's already in your pantry - ideal for long-term food utilization!

## Clean-up Convenience with Minimal Meals

At times of limited resources and busy schedules, prioritizing cleanup convenience while still enjoying nutritious meals is of utmost importance. One-pot and one-pan meals provide an efficient solution by limiting the dishes used and streamlining cleanup processes; many such meals incorporate multiple ingredients cooked simultaneously in one pot or pan for quicker and simpler cleanup processes - saving both cooking time and dishwashing efforts!

## Benefits of One-Pot and One-Pan Meals

- Simplicity: One-pot and one-pan meals offer easy meal preparation by requiring few utensils or ingredients; these recipes make quick cooking possible!
- Time Saving: With minimal meal prep required and few dishes to wash afterward, one-pot or one-pan meals offer time-saving solutions for busy individuals or families on the move.
- Versatility: These meals allow for greater versatility when selecting ingredients, so you can utilize whatever supplies are on hand or need using up, thus decreasing food waste, and saving money!
- Nutritious Options: One-pot and one-pan meals offer nutritiously balanced options, providing proteins, vegetables, grains, and legumes in just one dish.
- Flavor Infusion: When cooking ingredients together in one pot or pan, flavors blend and intensify, creating rich and delectable dishes.

## Example One-Pot and One-Pan Meals

- One-Pot Pasta Primavera: Combine pasta, vegetables, broth, and seasonings in a large pot before simmering it until both pastas are done, and sauce has thickened sufficiently.
- Sheet Pan Chicken and Vegetables: Arrange chicken pieces, potatoes and assorted vegetables on a baking sheet and drizzle them with olive oil before seasoning as desired for maximum tenderness when roasting in the oven.
- Quinoa and Black Bean Skillet: Prepare this skillet meal using quinoa, black beans, diced tomatoes, corn kernels, spices and garnished with avocado slices, salsa sauce and cheese as desired for ultimate deliciousness!

- Vegetable Stir-Fry: Make an easy and quick Vegetable Stir-Fry by stir frying an assortment of vegetables, tofu/chicken/rice with soy sauce and other seasonings in a wok or large skillet using soy sauce for seasonings for a nutritious yet fast dinner option!
- Slow Cooker Chili: Combine ground meat or turkey, beans, tomatoes, onions, and spices in your slow cooker before setting on low for several hours to let all those flavors come together and create thick and hearty chili.

## Tips for Easy Cleanup

- Use Non-Stick Cookware: For an easier cleanup experience, choose nonstick pots and pans as they will prevent food from sticking and reduce food waste.
- Line Baking Sheets with Parchment Paper: Before roasting or baking food, line the sheets with parchment paper to help avoid food sticking while simplifying cleanup. Doing this can keep food from becoming stuck to your surfaces as well as reduce cleanup efforts afterward.
- Prep Your Ingredients in Advance: Prepping ingredients early can save time on both prep time and cleanup later, simplifying your cooking experience while cutting cleanup costs down significantly. Chopped vegetables, measured spices, and prepared ingredients should all be assembled prior to beginning their journey towards being finished recipes and dishes!
- Soak Pots and Pans: When finished cooking, fill your pots and pans with warm soapy water immediately to loosen food residue and make cleaning easier later.
- Clean as You Go: For optimal cleanup after eating, wash all utensils, cutting boards and cooking tools while the meal simmers or bakes in one pot/pan. This way you will limit cleanup afterward!

By including one-pot or one-pan meals into your meal planning repertoire, you can easily enjoy convenient, nutritional, and delectable dishes while simultaneously cutting back on time and effort needed for cleanup.

## Classic Chicken Noodle Soup

**Ingredients:**

chicken broth (or bouillon cubes), canned chicken breast and dried pasta such as egg noodles. Carrots, celery, onion, garlic cloves, salt and pepper should all be present, along with dried herbs such as thyme or parsley for flavor enhancement.

**Instructions:**

Sautee onions, carrots, celery, and garlic in oil until vegetables have softened before adding chicken broth, shredded chicken, dried pasta, and herbs to simmer until pasta has reached desired doneness and adjust seasoning according to taste.

## Tomato Basil Soup

**Ingredients:**

Canned tomatoes, onion, garlic cloves, vegetable, or chicken broth (or bouillon cubes), dried basil leaves and oregano from dried flowers as well as sugar (optionally), salt peppercorns and heavy cream may all be combined into this delicious soup recipe!

**Instructions:**

Sautee onions and garlic until translucent before adding canned tomatoes, broth, basil, oregano, sugar, salt, and pepper for simmering over 20-30 minutes before blending to achieve smooth consistency if desired and stirring in some heavy cream if desired.

## Bean and Pasta Soup

**Ingredients:**

canned kidney beans or white beans, dried pasta, canned tomatoes, onion, garlic cloves, vegetable, or chicken broth cubes (bouillon cubes), Italian seasoning or dried herbs of your choice (like dried Italian seasoning), salt, and pepper for flavor.

**Instructions:**

Sautee onions and garlic until softened before adding canned tomatoes, beans, broth, pasta, herbs and seasoning to simmer until pasta is fully cooked - adjust seasoning according to personal taste!

## Minestrone Soup

**Ingredients:**

Canned tomatoes, kidney, or cannellini beans (regular, kidney bean-type beans are great), onions, garlic cloves, carrots, and celery; as well as dried pasta are used. Furthermore, vegetable or chicken broth or bouillon cubes with dried oregano and basil dried herbs along with salt and pepper make an essential base to this delicious soup!

**Instructions:**

Sauté onions, carrots, celery, and garlic until softened before adding canned tomatoes, beans, broth, pasta, and herbs. Simmer until pasta is fully cooked then adjust seasoning as necessary to achieve your preferred result.

## Sustained Energy Mixes for Trail Running and Bars DIY

Energy levels are key when venturing outdoors or camping trips. Customizing trail mixes and energy bars allows for customization to your nutritional needs and flavor preferences, with options like nuts, seeds, dried fruits, and optional ingredients such as chocolate chips or coconut flakes being easily added for customized balance between carbohydrates, proteins, and healthy fats. When creating homemade energy bars blend dates nuts oats cocoa powder vanilla extract as desired to form firm bite-size snacks that provide long-term energy for hiking backpacking or any outdoor activity!

## Meal Ideas for Outdoor and Campfire Cooking

Outdoor cooking presents opportunities to prepare creative and flavorful meals using campfires, portable stoves, or grills. Foil packet meals allow you to layer seasoned meats, vegetables, and potatoes inside aluminum foil pouches before sealing tightly and cooking over hot coals or on a grill until tender. Skewer marinated meats and vegetables for effortless grilling over an open flame; wrap hot dogs or sausages in biscuit dough or crescent rolls and roast over the campfire for an interesting twist on classic campfire fare; don't forget ingredients to create delicious s'mores before dinner ends!

## Shelf-Stable Ingredients for Sweets and Desserts

Assembling desserts outdoors is made simpler when using shelf-stable ingredients that require little refrigeration, such as no-bake energy balls made from rolled oats, nut butter, honey, and mix-ins such as chocolate chips or dried fruit. Make granola bars using oatmeal honey nuts seeds press into pan then refrigerate until set; or create banana boats by slicing banana lengthwise then filling it with chocolate chips and marshmallows wrapped tightly in foil before heating over hot coals until everything melts into gooey gooeyness!

## Creative Recipes Utilizing Foraged and Wild Edibles

Foraging for wild edibles adds an exciting element to outdoor cooking and allows you to incorporate fresh local ingredients in your recipes. Discover delicious wild greens such as dandelion leaves, lamb's quarters, or purslane for use in salads or sautés; harvest edible mushrooms like morels or chanterelles for earthy flavoring in pasta dishes and risottos; use wild berries or fruit as ingredients in desserts jams and sauces that bring nature inside! Always properly identify wild edibles when foraging to avoid polluted areas or pesticide-affected areas! When foraging makes sure to do it safely away from pollution or pesticide-affected areas as they could pose hazards!

DIY energy mixes and bars, outdoor cooking techniques, shelf-stable ingredients for sweets and desserts, foraged edibles from nature and wild edibles can all elevate outdoor dining experiences while adding delicious yet nutritious and satisfying options to meals outdoors.

Fermenting and Pickling Produce for Long-term Storage Fermenting and pickling produce are great ways to preserve fresh produce for longer storage while adding unique flavors.

## Sauerkraut

Ingredients and Instructions for Cabbage Sauerkraut Ferment: Clean cabbage finely before placing it into a large bowl and seasoning it liberally with salt, then massage until you notice release of juices from its cell walls. Pack the cabbage tightly into an empty, sterilized jar by pressing down tightly against its base, submerging itself completely under liquid from its insides, cover loosely to protect from light exposure for at least 1-2 weeks depending on taste preference, then transfer into fridge storage after the fermentation has completed.

## Pickled Cucumbers

**Ingredients:**

cucumbers, dill, garlic cloves, black peppercorns, mustard seeds, and vinegar - plus water, salt, and sugar as optional ingredients.

**Instructions:**

Slice cucumbers into thick slices and pack tightly into clean, sterilized jars with dill, garlic cloves, peppercorns, and mustard seeds for flavor enhancement. In a saucepan combine vinegar, water, salt sugar together until fully dissolved before pouring hot brine onto cucumbers at room temperature before refrigerating to develop flavor further over time.

## Pickled Jalapenos

**Ingredients:**

jalapenos peppers, vinegar, water, salt, and sugar as well as garlic as well as optional coriander or peppercorns for adding spice.

**Instructions:**

Slice jalapenos into rings or leave them whole. Place in sterilized jars along with garlic and spices before heating vinegar, water, salt, and sugar together until combined into a brine mixture and pouring it over your jalapenos while leaving some space at the top for expansion as it cools at room temperature before refrigerating them a few days later for consumption!

Before filling your ferments or pickles in homemade preserved foodstuffs, be sure to properly sterilize your jars to eliminate any possibility of spoilage from unwanted bacteria. Enjoy making delicious, preserved goodies!

## Protein-Rich and Eco-Friendly Dishes

**Beans and lentils provide an essential source of protein and essential vitamins and nutrients in emergency situations or everyday cooking alike, being sustainable and shelf stable. Utilize black beans, kidney beans, chickpeas, and lentils in various dishes for an additional nutritious boost!** Create delicious bean-based soups or stews by simmering beans with vegetables, herbs, and spices in an aromatic broth. Create protein-packed salads by tossing cooked lentils or beans together with fresh vegetables, grains, and tangy vinaigrette dressing - ideal as main courses! For even more versatility use bean-based burgers or meatless chili as main courses! Legumes and lentils provide both sustenance and flavorful meals no matter the circumstance!

## Baking with Homemade Legumes and Lentils

Homemade baked goods benefit immensely from including legumes and lentils as part of their ingredients list, adding nutrition, texture, and flavor. When adding legumes or lentils into homemade bread recipes at home, use cooked and mashed beans/lentils instead of all or some flour in bread/muffin/pancake recipes; use cooked chickpeas/black beans into brownie or cookie batter for increased moisture/protein. Combine pureed legumes with whole wheat flour yeast seasonings before rising before shaping the dough into hearty loaves before baking into hearty loaves - home-baked products allow customization as well as providing extra nutritional boost! Home-baked items are great ways of providing customized goodies while providing additional nutritional boost while giving additional nutrition boost! Home-baked goods often include extra nutritional boost.

## Cooking Conservation Strategies:

At times of water scarcity or when conserving resources, developing recipes to minimize water usage is paramount. Opting for techniques such as steaming, sauteing and roasting which use minimal liquid will allow your meal to retain flavor without using up extra liquid resources. Select recipes that make use of braising or stewing techniques where ingredients cook in their own juices with just minimal extra liquid added - cutting back on additional water consumption while providing flavor and nutrition boosts to soups, sauces, or stocks. Reusing cooking water from boiling vegetables or pasta also gives soups, sauces, or stocks an edge! Consider investing in water-efficient kitchen appliances such as pressure and slow cookers that use less water when tenderizing ingredients effectively. By prioritizing conservation and mindful cooking practices, you can lower water usage without compromising flavor or nutrition.

## Pantry Essentials to Prepare International Cuisine

Making international cuisine from pantry staples provides an exciting, flavorful approach to meal planning during challenging times. Stock your pantry with versatile ingredients like rice, pasta, canned beans, and tomatoes; these serve as the basis of many global dishes! Experiment with spices, herbs and condiments found in international cuisine to add depth and complexity to your meals, such as curry powder, soy sauce, salsa, or coconut milk. Or create classic Italian pasta dishes using ingredients found in every pantry such as garlic cloves, olive oil and dried herbs - an exciting and innovative approach! Create delicious Spanish-style chickpea stew with canned chickpeas, tomatoes and smoked paprika; or explore Asian cuisine by stir frying rice noodles with vegetables, tofu, and homemade soy sauce/sesame oil/rice vinegar sauces - creating tasty global cuisine-inspired meals is easier than you think! By stocking your pantry well and welcoming global flavors you can easily produce satisfying and mouthwatering dishes inspired by global cuisines from all around the globe!

## Comfort Foods for Stressed Times

Comforting foods offer nourishment for body and soul during times of anxiety or uncertainty, providing much-needed sustenance while simultaneously soothing our souls. Comfort food favorites like macaroni and cheese, mashed potatoes or chicken soup provide that familiar warmth we crave during this uncertain time in life. Create classic comfort dishes like meatloaf, pot pie or lasagna ahead of time and then reheat as needed throughout the week for easy meals and comforting snacks. Indulge in sweet treats like chocolate chip cookies, banana bread or warm apple crisp for dessert or snacks to complete this comforting menu! Try international comfort dishes such as ramen noodles, curry, or risotto to find moments of respite and enjoyment during challenging times. By including comfort foods in your diet rotation plan, they may provide moments of respite.

## DIY Project: Survival Cuisine: The Prepper's Recipe Collection

**Materials:**

- Long-lasting pantry staples (e.g., canned goods, dried beans, rice)
- Vacuum-sealed or freeze-dried foods (e.g., dehydrated fruits and vegetables, powdered milk)
- Shelf-stable condiments and seasonings (e.g., salt, pepper, spices)
- Cooking oil or fat (e.g., olive oil, coconut oil, lard)
- Whole grains (e.g., oats, quinoa, barley)
- Dried herbs and spices for flavoring dishes
- Canned protein sources (e.g., tuna, chicken, beans)
- Nutritional supplements (e.g., protein powder, vitamins)
- Portable cooking equipment (e.g., camping stove, portable grill)
- Fire-starting materials (e.g., matches, lighters)
- Cooking utensils (e.g., pots, pans, spatula)

- Water purification tablets or drops (for cooking water)
- Dehydrated soup mixes or bouillon cubes
- High-energy snacks (e.g., nuts, trail mix, granola bars)
- Emergency ration bars or meal replacement bars
- Ready-to-eat canned meals (e.g., chili, stew)
- Survival gardening seeds (for long-term food production)
- Food preservation supplies (e.g., canning jars, vacuum sealer)
- Fishing gear or traps (for obtaining fresh protein)
- Recipe books or printed recipe cards for reference

**Procedure:**

- Preparing your pantry for any situation begins by stocking long-lasting pantry essentials that provide essential nutrition and calories in an emergency. Start by organizing your shelves so supplies can easily be reached when they're needed.
- Include canned goods, dried beans, and rice as the foundation items in your survival cuisine plan. They all offer long shelf life and provide you with meals for a range of meals that you might prepare in an emergency.
- **Consider investing in vacuum-packed or freeze-dried foods, like dehydrated fruits and vegetables, powdered milk powdered meals or instant meals that remain lightweight yet retain nutritional value** for extended periods. These items offer convenient space savings while remaining lightweight with extended nutritional retention value.
- Don't forget to stock up on shelf-stable condiments and seasonings like salt, pepper, and spices to add extra flavor to your dishes. Olive oil, coconut oil or lard is essential when it comes to sautéing and frying - be sure to have plenty of it available as soon as you start sauteing and frying!
- Add whole grains like oatmeal, quinoa and barley for meals packed with energy-sustaining carbohydrates that will sustain you through to dinnertime. Dry herbs and spices will enhance the flavor and add variety to your diet.
- Make sure your diet includes enough protein - such as canned tuna, chicken, and beans - for muscle maintenance and repair. Add supplements like protein powder or vitamins, as necessary.
- To facilitate meal prep, it's essential to bring portable cooking equipment such as a camping stove or portable grill along with fire starting materials like matches or lighters. Furthermore, basic pots, pans and spatulas will prove indispensable during preparation of food items.
- Use water purification tablets or drops to ensure cooking water remains safe for consumption during a crisis and keep dehydrated soup mixes or bouillon cubes on hand to add flavor.
- Add energy-packed snacks such as nuts, trail mix and granola bars to your pantry for quick energy boosts, emergency ration bars or meal replacement bars can provide quick meal replacement when cooking is not feasible.
- Ready-to-eat canned meals such as chili and stew provide quick and convenient meal solutions with little preparation required, while investment in survival gardening seeds could enable long-term sustainability of food supplies.

- Stockpiling food preservation supplies like canning jars and vacuum sealers is key for protecting any excess food that might spoil. Furthermore, fishing gear or traps may come in handy when extracting fresh protein sources nearby.
- Assemble a list of recipes from recipe books or printed recipe cards as references for future meals using your survival cuisine supplies. Experiment with different ingredients and cooking methods until nourishing meals can be created using survival cuisine supplies.
- With these steps and supplies in your pantry, you will be well-equipped to create delicious yet nutritional meals during any crisis or emergency.

# BOOK 8: FIRE MASTERY: IGNITE SURVIVAL ART

Fire Survival is essential to ensuring safety in outdoor environments or emergency scenarios, providing warmth, light, and food preparation capabilities - making fire an invaluable survival asset in wilderness or disaster settings. Fire can present numerous risks when not managed effectively, such as burns and wildfires that cause extensive property damage. By understanding fire safety principles and adopting them accordingly, individuals can mitigate hazards, prevent accidents, and utilize fire safely for warmth, cooking, signaling and protection purposes. Educating themselves about fire survival techniques such as starting methods, behavior modification techniques and suppression allows individuals to navigate outdoor environments safely as well as respond effectively in case of emergencies involving fire.

## Principles for Lighting Fires Easily

Mastery of basic fire-starting principles is key for starting fires safely in outdoor settings or survival situations. Key principles include:

- Tinder: To start a fire quickly and sustainably, begin by lighting an initial spark using fine materials such as birch bark, cotton balls or dry grass. Kindling: Gradually add lightweight materials like branches or pine needles as kindling to feed and sustain your initial flame and create an ongoing fire.
- Fuel: Once the fire is lit, gradually add larger pieces of fuel such as logs or branches to keep it going **over time**.
- Airflow: For proper airflow to reach the fire and promote combustion, arrange materials into loose pyramid-shaped formation to allow oxygen into its path and promote proper combustion.
- Ignition: Use an ignition source such as matches, lighters or fire starters to ignite the initial flame and light your tinder pile. By following these principles and following safe practices when starting fires outdoors in various environments and climates, individuals can effectively start fires for comfort, safety and survival in various outdoor settings and conditions.

## Essential Tools and Equipment for Fire-Making

An essential tool and equipment needed for creating fire in outdoor settings or survival scenarios are:

- Ignition Sources: Always carry multiple reliable ignition sources like waterproof matches, lighters or ferrocerium rods as redundancy and reliability are essential when starting fires.
- Tinder Materials: When packing fire starters to use under challenging circumstances, lightweight yet easily ignitable tinder materials like cotton balls, dry leaves or commercial fire starters are key for effective ignition of any potential flames.
- Kindling Materials: Collect small and dry kindling materials like dry branches, pine needles or small sticks to start your fire slowly and feed its initial flame.
- Fuel: To fuel and sustain a fire over time, amass an ample supply of small sticks, larger branches, and logs in various sizes for fueling your fire.
- Fire Container/Pit: For fire safety in areas with flammable vegetation or strict environmental regulations, carrying a portable fire pit or container with you is highly advised. This way, any accidental sparks from starting up could be contained and managed safely and in turn managed more easily by professionals.

Individuals who assemble an effective fire-making kit can quickly start and manage fires in various outdoor settings and emergency scenarios.

## Items Essential to Assembling a Fire-Making Kit

Making an effective fire-making kit involves selecting essential items carefully to be fully prepared and reliable in starting fires. Items necessary in such kits may include:

- Waterproof Matches: Guarantee reliable ignition in wet or damp conditions.
- Lighter: Provide an easily transportable source of ignition for starting fires. Fire Starters: To help ensure fire ignition even during adverse conditions.
- Knives or Multi-Tools: For processing firewood, carving tinder, or creating fire-starting tools like feather sticks. Tinder Material: Pack lightweight yet easily ignitable tinder materials such as cotton balls, dried leaves, or wax-coated fire starters to provide fast and reliable fire ignition.
- Emergency Fire Blanket: As an additional safety measure, these blankets offer extra protection when extinguishing small fires or dousing flames during emergencies.

By compiling these essential items into an easily transportable fire-making kit, individuals can improve their ability to start fires quickly while staying prepared for outdoor adventures or emergency scenarios. Finding Tinder and Preparing Techniques:

Finding and preparing tinder effectively are essential steps in starting a fire in difficult circumstances, particularly under difficult circumstances. Tinder refers to easily ignitable materials which catch flame quickly such as dry leaves, grass blades, bark shavings or cotton balls coated in petroleum jelly; when searching for it gather fine and dry materials that ignite easily with sparks or flames. Fluffing or shredding tinder to increase surface area can improve its ability to catch flame as can arranging it loosely into bundles or nests for ignition and airflow; consider alternative sources such as Birch bark dry moss or newspaper as alternatives if traditional sources become scarcer or in cases such as this situation.

## Materials and Kindling

Selecting suitable materials and kindling are vitally important when creating a fire that produces sustained heat and flames. Kindling are small combustible items like twigs, small branches or split wood pieces which ignite quickly to provide initial fuel for a fire's ignition. Choose lightweight kindling that snaps easily and burns efficiently for effective fire starting, such as dry branches that easily snap. As your fire grows larger, gradually increases its size from small twigs and smaller sticks and branches up until bigger sticks and branches. When selecting kindling materials from dead standing trees or the ground itself, store it dry until needed for fire starting.

## Lays: Teepee, Lean-to, etc. Cabin/Log

As you build a fire lay - an arrangement of firewood - it is important to consider different configurations to optimize airflow, fuel consumption and heat distribution. Common designs of this sort include the teepee, lean-to, log cabin or pyramid designs. A teepee lay involves stacking kindling and fuelwood in an arrangement resembling an inverted cone shape with the tinder bundle in its center for early ignition, while lean-to lay involves leaning larger fuelwood against one piece of kindling to create an enclosed space where fire may spread unimpeded. Log cabin fire layouts utilize alternate layers of fuelwood stacked into either a square or rectangular pattern with tinder and kindling placed at its center and may include other materials or weather considerations that affect how quickly a fire starts up. Experiment with various firefly designs until finding one which best meets your individual fire-starting situation such as available materials, weather conditions or desired burn time.

## Starting Fire in Challenging Conditions

Starting a fire under challenging circumstances such as wet or windy conditions requires additional planning, patience, and creativity. For optimal fire-starting conditions in wet environments such as under overhanging rocks or fallen trees a dry spot could provide shelter - be prepared. Use natural fire starters such as dry pinecones, birch bark and fatwood resin as natural fire starters to quickly light damp tinder and kindling more easily. Build a platform out of dry rocks, logs, or green branches to elevate your fire off wet ground surfaces to prevent moisture absorption by your fire. When facing windy conditions, protect the fire by creating a windbreak or using your body as a shield against gusts of air. Consider windproof lighters/matches/compact wind screens as protection measures to keep its flame from extinguishment. Adapt your fire-starting strategies accordingly to effectively build and sustain fires for warmth, cooking or signaling purposes.

## Starting From Scratch

Natural fire starters can be invaluable tools in adverse weather or when dealing with damp materials, like **pinecone resin, dried grasses**, birch bark or fatwood from resin-rich tree stumps. Look out for readily available natural materials with flammable properties - like pinecone resin, grasses dry enough for mowing or resinous pinecone resin for example! Assemble and prepare natural fire starters in advance, keeping them dry and easily accessible when necessary. Crumble or shred fibrous materials, like dried grass or bark, to form tinder bundles which ignite quickly when struck by spark or flame. Use natural fire starters as an aid when traditional materials for starting fires become scarce or wet, such as when conventional kindling becomes unavailable or damp. Experiment with various natural starters to find those most capable of lighting flames under various weather and environmental conditions.

## Starting Fire DIY Resin, Pine

DIY fire starters crafted from resinous pine or other natural materials provide reliable ignition sources in challenging circumstances. Collect resin-rich pinecones, sap, or pitch from coniferous trees like pine, spruce, or fir to use in creating DIY starters for starting fires in these situations. Prepare the resin into manageable chunks by slowly heating it over an open flame or in an enclosed vessel placed over hot coals until it melts and flows freely. Make homemade fire starters by pouring molten resin into molds or pouring onto absorbent materials like cotton balls, wood shavings or dried grasses and waiting until it cools and solidifies completely before storing or using. To use, place a resin fire starter beneath a bundle of tinder and light it with spark, match or lighter to initiate combustion and spark larger kindling and fuelwood. Experiment with different resin sources and formulations until you find one tailored to your fire-starting needs and preferences.

## Considerations for Fire Safety

Safety should always be your number-one concern when working with fires, whether in an enclosed environment such as a fireplace or campfire or during an emergency. A few key considerations for fire safety:

- Clearance: Always leave enough clearance around a fire to minimize its spread to nearby objects or vegetation, including nearby buildings or vegetation.
- Supervision: Never leave an unattended fire unsupervised - always be present when children are near fires as this provides supervision from someone responsible.
- Extinguishing: Have an effective means for extinguishing fire nearby, such as water or a fire extinguisher and know how to use it effectively. Fuel Selection: Only select appropriate fuels for your fire and avoid materials that produce excessive smoke, sparks or fumes that might create hazardous situations.
- Wind Conditions: Always pay close attention to wind conditions as gusts of air may quickly spread embers and flames, creating an immediate danger. Proper Ventilation: Always ensure proper ventilation when using fire indoors to prevent carbon monoxide and other toxic gases from building up inside a room, which would otherwise trap carbon monoxide gas inside of it and potentially poisonous gases such as smoke from becoming trapped therein.
- Emergency Response: When responding to emergencies, always have an evacuation route and communication strategy in place. By prioritizing safety concerns over convenience when responding to fires, accidents can be reduced significantly, and injuries avoided.

## Maintain Safe Fire-Building and Management Behavior

Safe fire building and management practices should be integral parts of anyone working with fires for cooking, heating, or recreation purposes. Here are a few best practices you should be following when building or using fires:

- Start Small: To keep flames under control and the risk of uncontrollable fire, begin with a small fire that gradually adds fuel, gradually controlling both its size and intensity of flames. Steady Supervision: Maintain constant supervision over your fire to avoid accidents or uncontrolled flames occurring unknowingly.
- Proper Extinguishing: Once finished with your fire, douse it with water to completely extinguish it before stirring up its ashes to ensure all embers have died down. Comply with Local Regulations / Restrictions: Please abide by any local rules regarding open fires, burning bans and designated fire areas when using open flames and burning bans.
- Leave No Trace Principles: Apply Leave No Trace principles by minimizing your fire's environmental footprint, leaving the site as it was before starting it. Educate Others: Share fire safety knowledge and best practices with others to promote responsible fire behavior and avoid accidents.

By adhering to responsible behavior when approaching fires, you can enjoy their benefits while mitigating risks to yourself, others, and the environment.

## Understanding Open Flame Cooking Techniques

Open flame campfire cooking techniques provide an exciting and unique way to prepare meals outside. Some popular methods for doing this are:

- Grilling: Place food directly over flames or coals using a grill grate or other suitable surface, turning, or flipping as necessary to ensure even heating without overcooking or burning the food.

- Skewering: Thread food items onto skewers or sticks and place them over an open fire for cooking, which works great for items like kebabs, hot dogs, marshmallows, and vegetables.
- Foil Packets: When creating foil packets to cook food on the grill, fill each packet with seasonings, herbs, and ingredients such as seasoning salt. Place near or on a grate near coals for cooking - these versatile packs can accommodate anything from meats and seafood to vegetables and potatoes!
- Dutch Oven Cooking: For even heating distribution and consistent results when making soups, stews, roasts, or baked goods over the campfire use a cast iron Dutch oven with even heat distribution; arrange hot coals both atop and underneath to ensure even heat distribution and even cooking temperatures throughout.
- Campfire Cooking Tools: Make cooking on an open flame easier and more enjoyable by investing in campfire cooking tools like grilling baskets, tripod grills, cast iron skillets and pie irons to expand your culinary possibilities.

Experiment with different open flame cooking techniques to craft delectable meals and unforgettable dining experiences in nature.

## Rescue using Fire for Signaling

Fire can be an invaluable ally in an emergency, serving to alert rescuers of imminent danger and get people's attention quickly. Here are a few techniques for using it for signaling:

- Building a Signal Fire: Construct a signal fire using green vegetation or materials which produce thick smoke by stacking these items tightly into an assembly, then lighting them off to produce visible smoke columns.
- Light Pattern Creation: Use fire to generate Morse code signals or light patterns as a form of communication with rescuers, such as placing burning sticks or torches in specific patterns to convey messages or draw attention.
- Flares or Signal Fires for Emergencies: Keep flares or signal fires handy as part of your emergency kit to use as signaling in case of an emergency. Light them outdoors away from any materials which might ignite, for optimal visibility and efficacy.
- Contrast Creation: Position yourself and your signal fire in an area with high visibility, such as an open clearing or hilltop, to increase its chances of being noticed by rescuers. Make it stand out even more effectively against its surrounding landscape by placing your fire contrastingly against it.

By understanding and practicing this fire signaling techniques, you can increase the odds that emergency services can locate and rescue you in situations with limited visibility. Starting Fires in Urban Environments:

Starting a fire in urban environments requires careful consideration of safety, legality, and environmental impacts. Open fires may be restricted in some urban areas due to the risk of property damage and fire risks; however, in certain circumstances such as camping in designated areas or emergencies it may become necessary.

Before trying to start a fire in an urban environment, first check local regulations and obtain any required permission or permits. Next, choose a location free from flammable materials, buildings, and vegetation where there will not be an accidental ignition of uncontrollable fires, such as using a firepit, barbecue grill or designated fire area; use appropriate fire-starting materials like fire starters, kindling and small pieces of firewood instead of accelerants like gasoline or lighter fluid which could ignite unruly flames that spread rapidly outward.

Once your fire is lit, keep an eye on it closely to ensure that it stays under control and under your supervision. Prepare an extinguishing solution such as bucket of water or fire extinguisher nearby that you could quickly turn against it should any sparks arise; once finished with your burning session be sure to extinguish thoroughly by dousing with water while stirring the ashes thoroughly so no embers remain burning!

## Extending and Maintaining Fire Safety

Follow these tips for increasing the lifespan and maintaining safety when operating a fire:

- Fuel Management: Add fuel gradually, rather than all at once, to sustain an even flame and avoid overwhelming it with too much of it at one time. Oxygen Control: Manage airflow into and around the fire by adding or withdrawing fuel and altering ventilation accordingly to achieve optimum combustion conditions and ensure an even flame.
- Heat Distribution: When placing fuel into an ideal setup for combustion, arrange in such a way to maximize even heating without hot spots or uneven burning. Ash Removal: Regularly remove any ash or debris that accumulates around your fire to help ensure airflow remains unobstructed for proper functioning and provide proper airflow around its source of combustion.
- Safety Monitoring: Constantly monitor your fire for signs of overheating, flare-ups or spreading beyond its designated area and take appropriate actions as soon as necessary if any problems arise.
- Extinguish: Once the fire has died down completely, extinguish it completely by dousing it with water and stirring its ashes until they feel cool to touch.

By following these guidelines, you can safely maintain a fire for extended periods while decreasing risks related to accidents or uncontrolled spreading.

## Practice Makes Perfect: Building Fire Proficiency

Developing skills in starting and managing fires requires regular practice and hands-on experience. Master basic fire-starting techniques using various methods (friction-based methods, fire starters or matches and lighters). Also experiment with various kinds of tinder, kindling or fuelwood samples to understand their properties and behaviors under various conditions.

Develop adaptability and problem-solving abilities through practice by building and tending fires in various environments, weather conditions, settings, and scenarios to build fires with limited resources or adverse circumstances - simulating real world scenarios! Take the opportunity to learn from experienced practitioners like survival experts, outdoor guides, or firefighters as you incorporate their tips and techniques into your fire building practice.

With consistent practice and dedication, it is possible to develop confidence and proficiency in starting and controlling fires safely and efficiently, providing yourself with an essential skill for outdoor adventures, emergencies, or survival situations.

## DIY Project: Ignite Survival: Mastering the Art of Fire Making

**Materials:**

- Tinder materials (e.g., dry leaves, grass, bark)
- Kindling (e.g., small twigs, dry branches)
- Firewood (e.g., larger branches, logs)
- Fire starter materials (e.g., cotton balls, dryer lint, wax)
- Waterproof matches or stormproof lighters
- Ferrocerium rod or magnesium fire starter
- Char cloth or charred cotton fabric.
- Fire piston or bow drill kit.
- Tin can or metal container (for char cloth production)
- Knife or multi-tool.
- Fireproof container or fire pit
- Windproof lighter or torch
- Fire accelerants (e.g., petroleum jelly-soaked cotton balls)
- Fireproof gloves or heat-resistant gloves
- **Fire extinguisher or water bucket (for safety)**
- **Fireproof tarp or mat (for protecting the ground)**
- Firewood saw or axe (for processing larger logs)
- Fire poker or stick (for adjusting the fire)
- Fireproof cooking grate or tripod (for cooking over the fire)
- Fire-making guidebook or manual.

**Procedure:**

- Mastering fire making is an indispensable skill for survivalists and outdoor enthusiasts, following these steps can ensure safe and successful ignition of your flame.
- Gather dry leaves, grasses, and bark as tinder materials for starting your fire. Next gather kindling (small branches with dry needles or dried out branches), small twigs and kindling to add fuel for an initial spark and kindling (small sticks used as spark plugs), kindling logs as kindle for initial flames before lighting larger firewood logs to sustain your new flames once lit.
- Prepare cotton balls, dryer lint, or wax to spark the fire and easily ignite it. Make sure these materials are dry and readily combustible; commercial fire starters or homemade options may also work effectively as fire starters.
- Select an ignition method suited to your tastes and skill level; examples may include waterproof matches, stormproof lighters, ferrocerium rods or magnesium fire starters. Familiarize yourself with each method prior to needing it in an emergency. Practice using them regularly!
- For use with either a fire piston or bow drill kit, prepare char cloth or charred cotton fabric as an ember-catching material. Cut small pieces from cotton fabric and heat in an aluminum can or metal container until their blackened surfaces turn black without igniting into flame. Store these charred materials in waterproof containers until needed later.

- Create a fire structure out of the materials gathered. Begin with a small pile of tinder in the center, followed by kindling arranged like a teepee or log cabin formation and finally larger logs for airflow around your structure.
- Use your chosen ignition method to light tinder. Holding the flame close, allow the fire to catch before gradually adding more kindling as required and larger logs to maintain it.
- Once the fire has taken hold, add more fuel as necessary to sustain its flames. Use a fire poker or stick to adjust log positions for proper airflow; carefully monitor this flame until extinguishing entirely when done.
- Be cautious when lighting fires outdoors and always have an emergency fire extinguisher or water bucket handy, using fireproof gloves when handling hot materials, and placing a fireproof tarpaulin or mat under your fire pit as protection from heat damage to prevent an accidental start-up of an uncontrolled blaze.
- When cooking on an open flame, place a fireproof cooking grate or tripod above the flames and use pots, pans, and skewers directly over them - or for slower-cooking meals use a Dutch oven!

Develop and perfect your fire-making techniques through trial-and-error using various materials and techniques, consulting an authoritative guidebook or manual for extra tips and strategies, or making use of any available online videos and guides that offer expert instruction in fire making techniques.

# BOOK 9: PREPPER'S HERBAL MEDICINE HANDBOOK

Prepping is all about self-sufficiency and herbal medicine is an integral component to that goal. Utilizing plant medicine's healing powers for common ailments or injuries as well as overall wellness maintenance needs makes herbal remedies invaluable additions to a prepper's toolbox. By understanding its benefits and limitations as well as learning to identify harvest and prepare medicinal plants properly, preppers can build a comprehensive herbal medicine cabinet designed to support both everyday life and emergency situations.

## Benefits and Limitations of Herbal Remedies

Herbal remedies offer many advantages for health and wellbeing, including accessibility, affordability, and a comprehensive approach. Many medicinal plants contain potent bioactive compounds with anti-inflammatory, antimicrobial and antioxidant effects to support natural body healing processes and allow personalized tailored treatments based on individual preferences and needs.

However, it's essential to acknowledge the limitations of herbal remedies, including their variable potency and effectiveness; potential interactions between herbal treatments and medications; dosage requirements for proper administration; as well as potential interactions with medications that interfere with them. Although herbal medicine is used alongside conventional medical care to address serious or life-threatening conditions. Preppers must educate themselves on proper usage as well as limitations for herbal products and **consult healthcare providers when needed**.

## Prepper's Herbal Medicine Cabinet Construction

Building an herbal medicine cabinet requires collecting medicinal plants, supplements, and natural treatments that address common medical concerns as well as emergencies. Some essential items for such an arsenal could include:

Dried herbs for teas, tinctures, and poultices such as chamomile, peppermint and echinacea; essential oils with antimicrobial and soothing properties, like lavender tea tree oil or eucalyptus oil.

- Herbal supplements for immune support, stress relief and overall wellness include elderberry syrup, garlic capsules and ashwagandha extract. First aid herbs used for wound care and pain relief such as comfrey, plantain and arnica can also provide effective remedies.
- Reference books or guides on herbal medicine, foraging techniques and plant identification can increase knowledge and build skills.
- By creating an herbal medicine cabinet, preppers can prepare themselves to address various health needs and emergencies while building self-reliance and resilience.

## Identification and Harvest of Wild Medicinal Plants

Preppers seeking herbal remedies from nature will find learning the art of foraging medicinal plants invaluable. Start by familiarizing yourself with local plant species, habitats, and seasons for optimal foraging opportunities; utilize field guides, online resources as well as hands-on guidance from experienced foragers or herbalists when it comes to identifying medicinal plants in nature.

Harvest wild medicinal plants sustainably to ensure long-term health and viability of plant populations. Only take what you need, leaving enough behind for growth and reproduction in future years. When foraging on private property or wildlife habitat, always ask permission and utilize ethical foraging techniques which respect wildlife habitat and reduce environmental impact.

After harvesting medicinal plants for storage and use, ensure they are processed and prepared accordingly, such as drying herbs for teas and tinctures; infusing oils with topical applications for topical applications; or preserving in vinegar or alcohol-based extracts. By responsibly harvesting wild medicinal plants for medicinal use, preppers gain access to an assortment of natural treatments while forging deeper connections to land resources and protecting biodiversity.

## Herbal First Aid for Common Injuries and Ailments

Herbal first aid can provide natural alternatives for treating common injuries and ailments, providing effective ways of relieving pain without using standard medical interventions. Some key herbal treatments used as first aid include:

- Arnica can help reduce pain, inflammation, and bruising associated with sprains, strains, or minor injuries when applied topically as cream or ointment. consul Calendula can provide soothing comfort in cases of cuts scrapes burns by speeding healing time while alleviating inflammation.
- Aloe Vera: Aloe vera leaf gel boasts anti-inflammatory and cooling properties, making it the perfect remedy for treating sunburns, minor burns, and skin irritations.
- Lavender Essential Oil: Lavender essential oil has both analgesic and antiseptic properties, making it useful in alleviating pain and speeding healing in minor cuts, insect bites and abrasions.
- Chamomile: Chamomile tea or compresses can soothe skin irritations, insect bites and minor allergic reactions while simultaneously decreasing inflammation and encouraging relaxation.
- By adding herbal remedies to your first aid kit, you can effectively address an assortment of injuries and ailments while encouraging natural healing and wellness.
- Herbal Tinctures, Infusions and Decoctions: How Can They Help?
- Tinctures, infusions, and decoctions are three popular herbal preparations used to extract medicinal benefits from herbs for both internal and external consumption. Here's how you can create each type of preparation:
- Tinctures: To make a tincture, combine dried or fresh herbs with high-proof alcohol such as vodka or rum in a glass jar and let steep for several weeks, occasionally shaking to promote extraction. Strain through cheesecloth or fine mesh sieve before storing tincture in dark glass bottle for long-term use.
- Infusions: Infusing herbs in hot water extracts their medicinal benefits, so to prepare an infusion bring water to boil then pour it directly over them in a heatproof container and cover tightly before steeping for 10-20 minutes in your heatproof vessel before straining out and drinking as tea!
- Decoctions: Decoctions are like infusions but involve simmering tougher plant parts like roots, bark, or seeds for 15-30 minutes at low heat before turning down to simmer. Strain the decoction for use either as tea or topically depending on its intended use.

Master these herbal preparation methods and you can craft powerful remedies that promote health and wellbeing.

## Herbal Antiseptics and Wound Care Solutions

Herbal antiseptics and wound care solutions offer natural alternatives to commercial antiseptic products for disinfecting minor wounds or injuries, including cleaning them effectively with herbal antiseptics or wound care solutions. Some effective herbal remedies for wound care may include:

- Tea Tree Oil: Tea tree oil contains powerful antimicrobial properties which help cleanse and disinfect wounds, cuts, and abrasions. Dilute it first in carrier oils like coconut or olive oil before applying directly onto skin surfaces.
- Echinacea tincture or extract can be applied directly to minor wounds to enhance immunity and speed healing.
- Plantain: Plantain leaves can be chewed or ground up into paste for application directly on wounds to relieve inflammation, reduce pain and accelerate tissue healing.
- Goldenseal: Goldenseal root powder or extract is known for its antibacterial properties and can be applied topically to wounds to help combat infection and promote healing.

To use herbal antiseptics and wound care solutions effectively, start by cleansing the affected area with mild soap and water before applying a natural remedy as appropriate. Cover any exposed wound with an antibiotic bandage to avoid further contamination of its environment.

## Herbal Remedies for Pain and Inflammation Management

Herbal remedies provide effective ways of alleviating discomfort and inflammation without experiencing adverse side effects from pharmaceutical medications. Some popular herbal treatments for pain management and inflammation are:

- Turmeric: Turmeric contains curcumin, an effective anti-inflammatory compound which may help ease arthritis pain and muscle soreness symptoms. Take turmeric supplements or incorporate fresh or powdered turmeric root or powder into meals as part of a healthier lifestyle plan.
- Ginger: With natural analgesic and anti-inflammatory properties, ginger provides relief for headaches, arthritis pain and menstrual cramps. Drink ginger tea or take supplements from fresh ginger root when needed for maximum effectiveness.
- Willow Bark: Willow bark contains salicin, an analgesic and anti-inflammatory compound like aspirin that provides fast pain relief from headaches, muscle ache and joint discomfort. Take willow bark supplements or make willow bark tea for pain management related to headaches, muscle soreness or joint discomfort.
- Capsaicin: Capsaicin is the active compound found in chili peppers that provides natural pain-relief properties to ease neuropathy and arthritis-related inflammation and pain, providing temporary relief in affected areas by applying capsaicin cream or gel topically to them. For the best results apply a capsaicin cream directly over affected area for maximum effect.
- Integrating herbal remedies into your pain management regime will allow you to effectively alleviate discomfort while supporting overall health and well-being. Herbal Treatments for Digestive Disorders and Food Poisoning:

- Herbal remedies provide safe and natural solutions to alleviate digestive disorders or food poisoning symptoms, such as digestive upset or foodborne illness. Some commonly utilized herbs for this purpose:
- Peppermint: Peppermint is well-known for its soothing properties in aiding digestive discomfort, providing temporary relief of symptoms like bloating, gas, or indigestion. Drink peppermint tea or take peppermint oil capsules regularly to feel better quickly!
- Ginger: Thanks to its anti-inflammatory and anti-nausea properties, ginger can provide relief from nausea, vomiting and motion sickness caused by food poisoning. Drink ginger tea or chew on some fresh ginger root pieces to relieve symptoms quickly and naturally.
- Chamomile: Chamomile has soothing properties to ease an upset stomach and reduce inflammation within the digestive tract, and drinking tea made with it or taking supplements designed specifically to promote digestive wellness can provide some much-needed comfort.
- Activated Charcoal: Activated charcoal can absorb toxins and chemicals found in food to treat food poisoning effectively. As soon as possible after ingestion of potentially unsafe items or water sources, activated charcoal capsules or tablets should be consumed immediately for the best results.

## Immune-Boosters and Herbal Preventative Solutions

Herbal immune boosters may strengthen your body's natural defenses while supporting overall health and wellness. Some effective herbs for increasing immunity include:

- Echinacea: Echinacea is an effective immune stimulant and may help both prevent and shorten the duration of colds and flu symptoms. Take regular Echinacea supplements or drink Echinacea tea regularly to enhance immunity function and boost your health.
- Astragalus: Astragalus is an adaptogenic herb which supports immune function while aiding your body to adapt to stress. Take astragalus supplements or drink astragalus tea regularly to strengthen both immunity and resilience.
- Elderberry: Elderberries contain many essential antioxidants and vitamins to improve immune health by strengthening immunity against viral infections and strengthening protection from them. Elderberry syrup or tea should be regularly consumed to strengthen your immunity. For best results, take elderberry supplements regularly in a form like Elderberry Syrup for maximum effects on immunity health.
- Garlic: Garlic boasts antimicrobial and immune-enhancing properties which can help combat infections while improving overall health. Add fresh garlic into your cooking or take garlic supplements for maximum immune-enhancing potential.

## Herbal Remedies for Respiratory Illnesses and Allergies

Herbal remedies may provide effective remedies to alleviate respiratory ailments and allergies by relieving inflammation, relieving congestion, and supporting respiratory health. A few effective herbs for these purposes include:

- Eucalyptus: Eucalyptus has decongestant and expectorant properties which may help relieve symptoms associated with respiratory illnesses like colds, flu, and bronchitis. You can inhale its essential oil or use chest rubs containing it to ease congestion and promote clear breathing.

- Licorice Root: Licorice roots contain anti-inflammatory and expectorant properties which may help relieve symptoms such as airway irritation or coughing, providing soothing comfort in an instant. Enjoy some licorice tea or take supplements of Licorice extract to find relief for respiratory distress symptoms.
- Nettle: Nettle contains natural antihistamine properties which can provide instantaneous relief of allergy-related symptoms like sneezing, itching, and congestion. Regular consumption of either tea made from, or supplements derived from Nettle can reduce allergic responses while supporting respiratory health and maintaining respiratory wellness.
- Thyme: Thyme has antimicrobial and expectorant properties which make it ideal for combatting respiratory infections while soothing coughs and congestion. You can drink tea containing thyme or inhale its essential oil to ease respiratory symptoms.

## DIY Project: Herbal Healing Salve

**Materials:**

- 1 cup of organic olive oil
- ¼ cup of dried calendula petals
- ¼ cup of dried comfrey leaves
- ¼ cup of dried plantain leaves
- ¼ cup of dried lavender flowers
- 2 tablespoons of beeswax pellets
- 2 tablespoons of shea butter
- 1 tablespoon of vitamin E oil
- 10 drops of tea tree essential oil
- 10 drops of lavender essential oil
- 10 drops of chamomile essential oil
- Double boiler or glass measuring cup and pot.
- Fine mesh strainer
- Cheesecloth
- Glass jars or tins for storage
- Labels and marker

**Procedure:**

- Beginning by prepping your dried herbs. If necessary, crush or grind dried calendula petals, comfrey leaves, plantain leaves and lavender flowers into smaller pieces to unleash more powerful healing benefits.
- Begin by setting up a makeshift double boiler by filling a pot with several inches of water and placing a glass measuring cup or heatproof bowl on top of it, adding olive oil, spices, and dried herbs into the upper portion of your double boiler.

- Heat the water in a pot over low to medium heat and allow the herbs to infuse their essences into the oil for 1 to 2 hours, occasionally stirring and keeping an eye on its level to ensure no one drops through completely.
- Once the herbs have infused with oil, remove the top part of your double boiler from heat and let the oil cool slightly before straining through a fine mesh strainer lined with cheesecloth to separate out plant material and extract as much oil from them as possible.
- Return the infusion oil to either a double boiler or clean pot set over low heat, add beeswax pellets and shea butter, stirring continuously until both ingredients have completely melted into and blended with the infusing oil.
- Once the beeswax and shea butter has melted, take it off the heat before stirring to combine vitamin E oil, tea tree essential oil, lavender essential oil, and chamomile essential oil evenly across.
- Carefully pour the hot herbal healing salve mixture into clean and sterilized glass jars or tins before allowing it to cool at room temperature before sealing them up with lids.
- Once cooled and solidified, label the herbal healing salve jars with their ingredients as well as date. Store them in an area free from direct sunlight and excessive heat.
- To use herbal healing salve, simply apply a small amount directly onto clean, dry skin when experiencing minor cuts, scrapes, bruises, insect bites, rashes, or skin irritations.

# BOOK 10: SELF-SUFFICIENT LIVING: OFF-GRID EMBRACE

Off-grid living refers to any lifestyle which operates independently from public utilities such as electricity, water, and sewage systems. This alternative way of life presents various advantages and challenges - one of the principal ones being greater independence from external resources, leading to increased self-sufficiency and sustainability; another significant perk being increased environmental consciousness due to reliance on renewable energies such as wind or solar and practice of conservation methods; furthermore it increases resilience against power outages or disruptions, providing preparedness against emergency situations.

Off-grid living poses several unique challenges that require careful thought and planning, most significantly the initial costs involved with setting up off-grid systems - which may vary according to factors like location and desired level of self-sufficiency - along with ongoing effort and expertise needed for maintenance purposes - energy generation, water provisioning, waste removal etc. must all be managed independently and individuals needing energy must manage this aspect themselves too! Furthermore, off-grid life may necessitate lifestyle modifications or sacrifices such as limited access to urban amenities or conveniences available through urban living spaces.

Understanding the intricacies of off-grid living is vital when contemplating this lifestyle choice, requiring in-depth research and assessments of personal needs, resources and preferences as well as considering other variables like climate and regulatory considerations to make informed decisions on whether it suits. Geographic location, climate conditions and natural resource availability play major factors when it comes to feasibility and suitability for off-grid living; also essential is investing time and energy acquiring necessary skillset and knowledge required to sustain an autonomous off-grid lifestyle. Although challenging at times; off-grid living offers unique opportunities that enable greater autonomy while environmental stewardship as well as resilience against an increasingly interdependent world.

## Location Selection for Off-Grid Living

Selecting an optimal location is paramount to successfully living off-grid. Many factors must be considered when selecting an acceptable site; firstly, access to essential resources like sunlight, wind and water is vital when designing off-grid energy and water systems; considering renewable sources like solar or wind power will allow reliable generation throughout the year; also, proximity to rivers lakes or groundwater reserves is vital in providing water self-sufficiency through rainwater harvesting or well drilling are critical aspects.

Location and climate play an equally critical role when considering off-grid living feasibility and comfort. Locations with milder climates and plenty of sunshine tend to be better suited to solar power production as well as year-round outdoor activities; regions prone to extreme climate conditions, like harsh winters or intense heatwaves may present additional planning needs and investment needs for insulation solutions, heating/cooling products or solutions such as passive cooling technology solutions. Accessibility considerations including road access as well as proximity of amenities or emergency services must also be factored into this decision-making process for practicality and safety in day-to-day living situations.

Furthermore, regulatory considerations and zoning restrictions can impact where one chooses for off-grid living. Certain areas have specific building codes, environmental regulations and land use restrictions governing off-grid properties that must be observed to comply with safety standards; prospective dwellers should research local laws to avoid legal complications as well as meet safety standards in compliance. Choosing an ideal spot requires carefully considering natural resources, climate conditions, accessibility factors as well as any applicable regulations to create sustainable yet comfortable off-grid lifestyle solutions.

## Sustainable Homes: Eco-Friendly Designs for Off-Grid Living

Sustainable homes reflect a harmonious blend of eco-friendly designs and off-grid living principles, striving to reduce environmental impact while increasing self-sufficiency. Care is taken when crafting sustainable homes to reduce resource consumption, harness renewable energy sources such as solar panels or wind turbines for power production, promote ecological balance and eliminate dependence upon traditional utility grids for electricity generation.

Eco-friendly home designs rely on natural building materials like adobe, rammed earth, bamboo, or reclaimed wood as one of their cornerstones, featuring low environmental impacts while offering superior insulation properties to reduce heating and cooling needs. Furthermore, sustainable homes often incorporate passive design techniques like strategic orientation, natural ventilation, or green roofs to optimize energy use while increasing indoor comfort levels.

Building off-grid adds another level of complexity to sustainable home building, necessitating careful planning and innovative solutions. Off-grid homes must include robust energy storage systems, efficient water management techniques and alternative waste disposal methods - while builders also must consider climate, topography and available resources when planning resilient dwellings that support themselves self-sustainably. Yet all the challenges can provide unparalleled freedom, resilience, and connection to nature!

## Off-Grid Food Production with Gardening and Permaculture

Off-grid living requires self-sufficiency in all areas, from housing to food production. Gardening and permaculture play important roles in off-grid food production by helping individuals cultivate nutritious produce without being dependent on external supply chains.

Gardening off-grid involves cultivating various fruits, vegetables, herbs, and edible plants using sustainable and organic practices. Off-grid gardeners employ techniques like companion planting, raised beds and mulching to maximize yields while conserving water resources. Permaculture principles guide off-grid garden design as they emphasize ecological harmony, diversity, and resilience - mimicking natural ecosystems while producing self-sufficient food forests where plants, animals and microorganisms cohabitate in mutualistic relationships.

Off-grid food production enables individuals to take control of their own food supply chain and ensure access to fresh, nutrient-rich produce year-round without dependence on industrial agriculture. Furthermore, off-grid gardening promotes deeper connections to nature while encouraging sustainable land stewardship practices - from rooftop gardens in urban settings to sprawling permaculture homesteads in rural settings, off-grid food production plays an essential part in creating resilient communities built upon self-reliance and resilience.

## Waste Management in Off-Grid Settings

Off-grid waste disposal environments present unique challenges and opportunities that require innovative approaches to minimize environmental impact while increasing sustainability. Dwellers living outside municipal services must develop alternative waste disposal plans which accommodate organic, recyclable, and non-recyclable waste streams effectively.

Composting is at the core of off-grid waste management, enabling residents to convert organic material from organic waste into nutrient-rich compost for soil enrichment. Off-grid composting toilets offer an eco-friendly alternative to conventional sewerage systems by turning human waste into compost through natural decomposition processes. Recycling and upcycling initiatives help minimize generation of non-recyclable trash as communities repurpose materials used during construction into construction materials or creative projects reusing materials for construction, crafts, or creative uses.

Off-grid waste management settings prioritize the principles of reduce, reuse, and recycle to foster resourcefulness and environmental stewardship. By adopting sustainable waste practices such as composting facilities or community recycling programs, off-grid communities minimize their ecological footprint while conserving natural resources while supporting circular economies. Such settings demonstrate sustainable living that can work harmoniously with nature.

## Transportation Solutions Off-Grid

Off-grid communities present unique transportation challenges due to remoteness and limited access to traditional infrastructure. Therefore, finding transportation solutions tailored to this lifestyle requires creativity, innovation, and dedication towards sustainability.

Off-grid transportation solutions often prioritize energy efficiency and self-sufficiency, with residents opting for eco-friendly modes like bicycles or electric vehicles (EVs) for transport - such as pedal carts. Solar charging stations may be installed strategically throughout an off-grid community to support electric vehicle usage while decreasing fossil fuel reliance.

Off-grid areas often lack reliable road infrastructure; therefore, alternative modes of transport such as off-road vehicles, ATVs or boats and canoes may be required for travel. Community transportation initiatives, like carpooling or shared vehicle programs may help maximize resource use while simultaneously decreasing environmental impact in such settings.

Transport solutions for off-grid living must take an inclusive approach that considers terrain, climate, community needs and resources available to them. By adopting innovative technologies and engaging in sustainable practices for transportation services, off-grid communities can maximize mobility while simultaneously decreasing their ecological impact and quality of life.

## Financial Planning for Off-Grid Living

Off-grid living requires careful financial planning that considers expenses, income sources and long-term sustainability. Although off-grid living provides opportunities to reduce utility bills and rely less on external services for savings purposes, it also necessitates significant upfront investments into infrastructure as well as ongoing maintenance requirements.

Financial planning for off-grid initiatives starts by conducting an in-depth assessment of initial expenses associated with land acquisition, construction costs, renewable energy systems, water management infrastructure and amenities essential to daily life. Next comes budgeting ongoing expenses related to maintenance repairs or upgrades as part of long-term financial stability planning.

Off-grid living can vary depending on individual circumstances and skill sets; income sources for off-grid residents depend heavily on those' circumstances and abilities. Some off-grid dwellers generate money via digital technologies like freelancing or consulting work remotely while others pursue alternative livelihoods like farming, artisanal crafts production, renewable energy consulting services or ecotourism in support of their off-grid lifestyle.

Sustainable financial planning for off-grid living involves diversifying income sources, managing debt efficiently and building emergency savings to be prepared for unexpected issues. Furthermore, off-grid residents may investigate bartering systems, community sharing projects or mutual aid networks as an additional form of aid that reduces their reliance on traditional monetary systems.

Off-grid living requires careful financial and resource management; yet its advantages include increased independence, self-reliance, and fulfillment. By adopting proactive approaches to their financial planning and living sustainably principles, off-grid residents can achieve increased financial resilience while fulfilling their **dreams of living off the grid**.

## Health Medical Care Off-Grid: Self-Sufficient Practices

Off-grid living requires residents to rely on themselves when it comes to healthcare services as access may be limited due to remote location limitations. Residents must prioritize preventive medicine, holistic wellness practices and emergency preparedness to maintain optimal health and well-being in this type of living situation.

Off-grid living requires health practices to take an integrated approach that encompasses physical, mental, emotional, and spiritual wellbeing. These may include regular exercise sessions and nutritious diet choices as well as stress management techniques, mindfulness practices and community support networks designed to promote overall well-being.

Off-grid residents must also be equipped to address medical emergencies and minor injuries with limited resources, ideally before professional help arrives - something which basic first aid training, wilderness medicine skills and knowledge of herbal remedies may assist with. Professional assistance could take hours or days away in these circumstances.

Off-grid communities tend to favor self-sufficient healthcare practices like herbalism, natural remedies, and alternative therapies as an adjunct to conventional medical approaches. Herb gardens, medicinal plant identification and DIY herbal preparation allow residents of off-grid communities to take control of their own healthcare while using nature resources to treat common ailments.

Though living off-grid may present its share of challenges, living this way offers unique opportunities to build resilience, self-reliance, and holistic wellness. By adopting self-sufficient health practices and cultivating an atmosphere of mutual support and empowerment in community living environments off grid can foster individual as well as collective wellbeing.

## DIY Project: Off-Grid Solar Power System
**Materials:**

- Solar panels
- Charge controller.
- Deep-cycle batteries
- Power inverter
- DC disconnects.
- Mounting hardware (brackets, bolts, etc.)
- Battery cables
- Wire connectors
- Circuit breakers
- Solar panel cables
- Junction box

**Procedure:**

- Assess your energy needs and establish what size of solar power system will best meet them. Calculate wattages of appliances and devices used, to estimate how many panels may be necessary.
- Pick an optimal location to install solar panels - one with full exposure to sunlight throughout the day - that allows maximum sunlight penetration without obstruction from trees or buildings that could cast shadows onto them.
- Secure solar panels to your roof using the provided mounting hardware, making sure they're angled correctly for maximum sunlight absorption.
- Connect solar panels and charge controllers using solar panel cables; the charge controller regulates electricity flowing from solar panels to avoid overcharging of batteries.
- Connect the charge controller to deep-cycle batteries using battery cables; these will store any energy generated from solar panels until needed later.
- Install a power inverter to turn DC power stored in batteries into AC current that's suitable for household appliances.
- Connect the power inverter to the batteries using appropriate wiring and connectors, installing circuit breakers to protect against overloads or short circuits in case they occur, as well as circuit breaker switches for protection of overloads or short circuits in case they arise.
- Install a DC disconnect switch between the batteries and inverter for safety measures, providing easy isolation during maintenance or emergencies. With such an arrangement in place, maintenance tasks and emergencies will no longer shut off all systems at once.
- Install a junction box to organize and protect wiring connections, making sure they're all secure and insulated for safety against electrical hazards.

- Once installed, make sure all parts of the system are functioning as planned by conducting comprehensive tests to make certain all aspects work as they should. Monitor its performance periodically and adjust as necessary to increase efficiency and ensure optimal functioning.
- Your off-grid solar power system enables you to reap the rewards of renewable energy while living an off-grid lifestyle and living independently.

# BOOK 11: HOME DEFENSE FOR PREPPERS

Evaluating vulnerabilities is an integral component of building security and preparedness in any home, as the process entails identifying any weak points where intruders could gain entry or threats could penetrate. Common weak spots include poorly secured doors and windows, darkened or isolated areas around the property and outdated or malfunctioning security systems - common weak points include poorly locked and secured doors/windows as well as malfunctioning security systems which allow intruders access. Regular assessments allow homeowners to prioritize areas need improvement as well as implement targeted measures which effectively mitigate risks effectively against threats effectively while adapting to changing circumstances while adapting and responding quickly when new vulnerabilities may present themselves.

## Home Security Systems and Surveillance Technology

Modern security systems provide advanced tools to defend against intruders while monitoring activity around your property. Modern systems may include features like motion sensors, door/window sensors, security cameras and alarm systems connected to monitoring services; surveillance technology enables homeowners to keep an eye on their property remotely as well as receive real-time alerts of suspicious activities as evidence in case of security breach; integrated with smart home technology allows seamless control and automation of security measures which enhance convenience while increasing effectiveness at protecting homes from intruders!

Reinforcing Doors, Windows and Entry Points Strengthening doors, windows and entry points is crucial to prevent unauthorized access and increase overall home security. Install high-quality locks, deadbolts, and strike plates on doors; secure bars or grilles on windows may help deter break-ins. Reinforcing techniques may include upgrading to impact-resistant doors and windows, installing shatterproof glass and adding layers of security such as door jammers or security film to fortify entry points against intruders and provide faster response or escape times in case of security threats.

## Establish a Strong Perimeter with Fences, Gates, and Barriers

A secure perimeter around your property is key for home security and deterrence. Fences, Gates, and barriers all play their parts to keep intruders out and reduce potential breaches in security. Installing physical barriers like fences, gates, walls, or hedges around a property to define its boundaries and control access points. An effective perimeter provides more than just security benefits: it acts as a deterrent against intruders while clearly marking property boundaries and providing added privacy and seclusion for residents. Selecting materials and designs for perimeter barriers depends on factors like security needs, aesthetic preferences, and environmental considerations. By investing in strong perimeter protection measures, homeowners can create a safe space in which their family can live.

## Establishing Safe Rooms and Emergency Shelters

Safe rooms and emergency shelters can serve as vital retreats in times of emergencies like home invasions, natural disasters, or any other forms of threat. Safe rooms are fortified spaces within a home that have been specifically constructed to withstand external forces and offer residents the security they require. These rooms may include reinforced doors, walls, and ceilings as well as communication devices and emergency supplies as well as provisions for ventilation and sanitation. Homeowners should also consider installing emergency shelters into their homes as additional protection from severe weather events and nuclear threats.

By having both safe rooms and underground bunkers for such protection in place, homeowners can increase resilience and preparedness against unexpected emergencies that arise within their household.

## Home Defense Weapons and Training

Weapons used for home protection should be an essential element of a comprehensive security plan for you and your loved ones. While weapon selection will depend upon personal preference, familiarity, legal considerations and any applicable policies or restrictions; common options include firearms, pepper spray or tasers as non-lethal options and even makeshift options such as baseball bats or knives that you improvise as weapons to defend yourself with when necessary.

Owning a weapon is only part of the solution; training and proficiency are vitally important to using it effectively in an emergency. Engaging professional instruction for firearm handling, self-defense techniques and situational awareness will allow homeowners to respond confidently and decisively in case of security threats.

Self-defense training extends far beyond physical techniques; it also covers mental preparedness and conflict de-escalation strategies. Adopting an attitude of vigilance, assertiveness, and calm under pressure will enhance your ability to protect both yourself and your home while simultaneously decreasing violent confrontations from becoming violent encounters.

## Establishing Neighborhood Watch Programs

Neighborhood watch programs are community-driven initiatives intended to prevent crime and enhance overall safety and security in residential areas. By mobilizing residents to be alert of suspicious activities in their immediate surroundings and look out for each other, neighborhood watch programs can serve to thwart criminals while creating strong senses of cohesion within communities.

Establishing a neighborhood watch program involves organizing meetings, recruiting volunteers, and working closely with law enforcement agencies in your community to formulate effective crime-prevention strategies. Participants in such programs often receive training in how to recognize suspicious behavior, implement security measures effectively and communicate efficiently with authorities.

Regular patrols, neighborhood meetings and communication channels such as social media groups or messaging apps allow members to share and collaborate information effectively. By working collaboratively to address common concerns and pool resources for protection of homes and neighborhoods, neighborhood watch programs empower residents to play an active part in protecting themselves and their neighborhood.

## Utilizing Guard Dogs for Home Protection

Guard dogs can provide invaluable home security, acting both as physical deterrents against intruders and as loyal companions for homeowners. Properly trained guard dogs can detect intruders, alert occupants to potential threats, apprehend trespassers through their presence or bark, detect intruders in real time and deter trespassers through presence and bark alone!

Selecting and training the appropriate breed and training regimen are critical components to optimizing home protection with guard dogs. Breeds known for their intelligence, loyalty and protective instincts like German

Shepherds, Rottweilers or Doberman Pinschers tend to make good candidates for guard dog training programs.

Training programs for guard dogs typically focus on obedience, aggression control and protective behaviors with particular emphasis placed upon socialization, desensitization to environmental stimuli and bonding with owners. Homeowners must provide necessary care, exercise, and mental stimulation to guarantee optimal success of these pets in protecting homes against intruders.

Protector dogs provide invaluable defenses against intrusion into homes when properly trained, giving residents peace of mind knowing there is always someone watching over their property.

## Develop Code Words and Signaling Systems

Code words and signaling systems provide a discreet yet effective means of communication among family, neighbors, and security authorities in times of emergency or security threats. Individuals can create code words as predetermined phrases or signals which convey specific instructions without alerting potential intruders of danger, thus helping alert others or coordinate responses without alerting potential intruders to danger or any responses needed.

Code words can be any simple phrases or numbers agreed upon beforehand by members of a household or community and only known to them. A code word such as "pineapple" could signify potential security threats to prompt everyone involved to implement predetermined safety measures or reach out for assistance immediately.

Signaling systems use visual or auditory cues to discreetly communicate messages or convey information. For instance, hand signals, light patterns or whistle codes may be employed to indicate distress, signal for assistance or coordinate responses between individuals without alerting potential threats.

Coding words and signaling systems with family or neighbors to prepare everyone for emergency or security incidents quickly is vitally important to being ready to act effectively in an emergency. By setting clear communication protocols and remaining alert in potentially risky environments, individuals can enhance their capacity for self-protection as well as others around them.

## Implementing Home Security Drills and Exercise

Implementing home security drills and exercises is critical to make sure all occupants of the household are prepared in case of an emergency or security threat. Regular drills help reinforce emergency protocols, familiarize residents with evacuation routes and assembly points and identify any weaknesses within your security plan that need improving.

Home security drills should include scenarios revolving around break-ins, fires, medical emergencies, and natural disasters so occupants are ready for potential threats that arise in various forms. Conducting drills at different times during the day or night helps simulate realistic conditions while testing individual's response in various circumstances.

Home security drills should include more than evacuation procedures and emergency responses; role-playing exercises that simulate interactions with potential intruders or emergency responders can also help build confidence, enhance communication skills, and ensure everyone knows their roles and responsibilities in an emergency scenario.

Reviewing and updating home security plans based on drill results is critical for maintaining readiness and effectiveness. By prioritizing preparedness training exercises and practicing resilience against threats to protect themselves and their property against challenges that arise, homeowners can increase resilience against adverse situations while building resilience that allows them to withstand potential dangers more readily.

## Maintaining Operational Security

Operational security, also referred to as OPSEC, is paramount in safeguarding sensitive information and mitigating vulnerabilities which could be exploited by potential threats. OPSEC involves identifying, controlling, and safeguarding information which could be exploited by adversaries to harm individuals, organizations, or assets.

OPSEC measures typically taken for home security include restricting the dissemination of personal information, being cautious when posting travel plans or daily routines on social media and safeguarding sensitive documents and electronic devices that could become targets for thieves and hackers.

**Behavioral OPSEC involves being mindful of conversations or behaviors which might inadvertently disclose details about one's home security measures, vulnerabilities, or valuables.** For instance, this might mean refraining from discussing security arrangements in public settings, as well as taking precautions against eavesdropping and surveillance attempts.

Maintaining OPSEC measures through constant review and updates in response to changing circumstances or threats is vitally important for effective security management. By prioritizing discretion, vigilance, and proactive risk management homeowners can reduce exposure to security risks while improving overall safety and security.

## Understanding Threat Levels and Response Protocols

Knowledge of threat levels and response protocols is vital in effectively evaluating and responding to security threats within a home environment. Threat levels depend on various factors including location, time of day, known vulnerabilities and current events and it's critical for homeowners to recognize and prioritize potential threats accordingly.

Response protocols provide guidelines on what should be done in response to various threats or emergencies, including how to assess a situation, notify others, and take necessary actions to reduce threats or reduce danger. Protocols should be tailored specifically for your household based on factors like the presence of children or older individuals, physical limitations, available resources.

Reviewing and practicing response protocols with all household residents regularly will ensure everyone is ready to respond effectively to emergencies or security threats in their own homes. This may involve conducting tabletop exercises simulating various situations, discussing roles and responsibilities among housemates, as well as pinpointing any areas for improvement within a response plan.

By understanding threat levels and response protocols and making sure everyone in their household can act swiftly when facing threats to themselves or property, homeowners can increase resilience against adverse situations while better protecting both themselves and their assets from future danger.

Fortifying Your Home against Natural Disasters Fortifying your home against natural disasters is vitally important, not only to minimize damage but to also ensure its occupants' wellbeing during extreme weather events or geological hazards. Fortification measures depend upon what types of natural hazards occur in your area - these measures could involve reinforcing structural elements, securing loose items, and creating safe zones within the house itself.

Fortification measures against hurricanes, tornadoes, or high winds include installing impact-resistant windows and doors; reinforcing roof structures; and securing outdoor furniture or structures which could become projectiles during strong winds. Furthermore, creating an indoor safe room or storm shelter gives occupants somewhere safe to take refuge during severe weather events.

Fortification measures in earthquake-prone regions typically focus on reinforcing building foundations, securing heavy furniture or appliances to prevent them from toppling over, installing automatic shut-off valves on gas lines to decrease fire risks or explosions, retrofitting older homes with modern seismic safety features to lessen potential damages, reduce collapse risks during an earthquake and minimize injury risks.

Fortification measures in areas prone to flooding or wildfire may involve elevating homes above flood levels, installing fireproof roofing materials and siding and creating defensible space around their properties by clearing away brush.

Maintaining and inspecting the fortification measures at your home on an ongoing basis ensures they remain effective and up to date, decreasing the risks of natural disaster damage or injury and strengthening resilience against extreme weather events or geological hazards. By taking proactive steps against natural disasters and fortifying your home against them, you can increase resilience against such impacts while strengthening resilience against extreme weather events or geological hazards.

## DIY Project: Home Fortress: Strengthening Your Defenses

**Materials:**

- Plywood sheets
- Nails or screws
- Metal brackets
- Door reinforcement kit
- Security window film
- Motion-activated outdoor lights
- Heavy-duty deadbolt locks
- Security cameras
- Alarm system.
- Door barricade bar
- Reinforced strike plates

- Security door chains
- Window bars or grilles
- Shrubbery or landscaping deterrents
- Safe room supplies (water, food, medical kit)

**Procedure:**

- Beginning by assessing your home's vulnerabilities. Consider areas like windows and doors as potential entryways into your residence.
- Add additional protection from forced entry by covering windows and glass doors with plywood sheets secured using nails or screws, reinforced with metal brackets to strengthen them further.
- Install a door reinforcement kit containing reinforced strike plates, longer screws and additional locking mechanisms on all your doors to increase security. Incorporate heavy-duty deadbolt locks and security chain door chains as additional measures of defense for maximum protection.
- Add security window film to windows for increased resistance against break-ins, making entry harder for intruders. This film helps hold glass together, so it becomes harder for intruders to gain entry.
- Install motion-activated outdoor lights around your home's perimeter to deter intruders from approaching without being seen and make life harder for potential intruders who attempt to infiltrate unnoticed. They will illuminate any movement outside, making it harder for intruders to approach undetected.
- Establish security cameras around your property to monitor activities and deter threats, placing cameras strategically such as at entryways or areas vulnerable to attacks.
- Consider investing in an alarm system with sensors on doors and windows that will alert you of any unauthorized entry attempts and deter potential intruders from breaking in. Such measures could reduce any unwelcome surprises when visitors arrive unexpectedly or help deter intruders from breaking in.
- Install door barricade bars to inward-opening doors to deter attempted forced entries and ensure you can securely close them against kicks, breaches and attempts at forced entry. They offer additional layers of defense by reinforcing against kicks or breaches while adding an additional measure against forced entries.
- Add window bars or grilles to ground-level windows for increased protection, making it harder for intruders to gain entry through these entry points. This may prevent access through windows by intruders who pose as genuine customers.
- Use shrubbery and landscaping deterrents to form an organic defense barrier around your home. Plant thorny bushes or dense foliage under windows as deterrents against intruders trying to enter through them.

Develop a safe room inside of your home stocked with supplies like water, food, and a medical kit in case of home invasion or another crisis. This room could serve as your last line of defense in such instances.

# BOOK 12: CAMPING WITH RVS

Finding an RV that meets all your survival needs involves considering various aspects, including size, durability, off-road capabilities, and self-sufficiency features. When selecting an RV model designed specifically for off-grid living - complete with robust construction features such as ample storage space and off-road tires capable of traversing challenging terrain - make a smart choice based on family or group size; prioritize functionality over luxury amenities; look out for one equipped with solar panels, water filtration systems or backup power sources to increase self-reliance during survival situations.

## Setting Up an RV for Off-Grid Living

Outfitting an RV for off-grid living requires installing essential systems and equipment designed to promote self-sufficiency and resilience. Install solar panels or wind turbines as a source of renewable energy; install water purification and filtration systems for drinking water quality improvement, propane tanks or diesel generators as backup power source, roof racks or storage boxes to maximize space; add locks, alarms or surveillance cameras as security features to provide added peace of mind in remote locations; consider additional storage solutions (roof racks/boxes etc.) until eventually reaching self-sufficiency is achieved.

## Management of Water in an RV

Water management in an RV requires careful and effective resource allocation. Implement water conservation practices like using low-flow fixtures, taking shorter showers, reusing graywater as non-potable supplies, and installing an integrated purification and filtration system that ensures safe drinking water sources such as rivers, lakes and rain collection systems are protected and treated correctly for drinking purposes. Monitor levels regularly while replenishing supplies, when necessary, to provide both adequate hydration and sanitation within your vehicle.

## Solar Panels and Generators

Generating power for RV off-grid living requires reliable and sustainable solutions. To harness solar energy for use by electrical appliances and devices, consider installing solar panels on the roof of your RV to collect it and recharge batteries from solar energy. Supplement solar with generators, wind turbines or fuel cells as backup solutions during periods with lower sunlight; further invest in energy-saving appliances and LED lighting to minimize power usage while maximizing system efficiency.

## Securing and Strengthening Your RV to Prevent Threats

Safety and security in an RV are of vital importance for survival purposes. To deter intruders and protect against potential threats, reinforce exterior doors, windows and locks using reinforced materials - then install security cameras, motion sensors and alarm systems as monitoring tools that alert to suspicious activities or entry attempts. Consider fortifying it further using barricades, window bars or bulletproof panels - additional defensive measures in high-risk environments might provide extra defense from unwanted entry attempts.

By carefully equipping and selecting an RV to meet survival purposes, you can increase your independence, safety, and comfort off-grid. Installing efficient water and power management systems along with robust security measures helps prepare yourself to face remote living challenges and thrive during survival situations.

## Mobile Communication Solutions for Remote Areas

Communication solutions in remote areas can help travelers and residents stay in contact and access emergency services, information, and support when traveling or living off-grid. While traditional cell coverage might be limited or unreliable in these environments, satellite phones, portable Wi-Fi hotspots, and satellite internet services all provide reliable options to connect.

Satellite phones offer global coverage that enables users to make calls, send text messages and access data from virtually anywhere regardless of cell coverage or infrastructure availability. Portable Wi-Fi hotspots utilize satellite or cellular networks for internet connectivity in remote areas so users can remain online via emails, messaging apps or social media and communicate easily using email, messaging apps or social networks.

- Emergency Beacons or Personal Locator Beacons (PLBs), along with mobile communication devices, serve as essential lifelines in remote regions by transmitting distress signals and GPS coordinates directly to emergency response officials in case of an emergency or crisis. They're particularly useful for outdoor enthusiasts, adventurers, and remote workers who might encounter emergencies while exploring or traveling across unfamiliar terrain.
- Investment in mobile communication solutions designed specifically to the unique challenges of **remote areas allows travelers and residents alike to stay in contact, access emergency assistance when required and experience peace of mind while exploring** nature or living off grid.
- Emergency Repair and Maintenance for RVs RV owners require emergency repair and maintenance skills to protect the safety, reliability and longevity of their vehicle when traveling or living on the road. From mechanical issues, electrical concerns, or structural damages; having knowledge and tools at their disposal is invaluable to avoid unexpected breakdowns that disrupt travel plans and cause costly breakdowns or disruptions of travel plans.
- Essential RV maintenance tasks involve regularly inspecting fluid levels, inspecting tires, and performing comprehensive examinations on key systems like brakes, suspension, and electrical components. Furthermore, learning to identify common leaks, appliance malfunctions or generator issues early can enable owners to address minor repairs before they escalate into larger ones.
- Be prepared for unexpected repairs or maintenance tasks on the road by stocking your RV with essential tools, spare parts, and emergency supplies such as tire repair kits, duct tape sealant spare fuses as well as basic hand tools to quickly address unexpected repairs or emergencies quickly and effectively.

With DIY repairs proving limited in scope and potential mechanical issues beyond DIY capabilities becoming an increasing threat, having access to professional assistance through roadside assistance or RV emergency service plans may offer added peace of mind and professional expertise when required. Many clubs and membership organizations provide customized programs specifically addressing RVers' needs with towing, repair services and emergency support available as needed.

Armed with the knowledge, tools, and resources necessary for emergency repairs and maintenance tasks on your RV, you can enjoy stress-free travel adventures on open roads.

## Stockpiling Essential Supplies in Limited Space

Storing essential supplies efficiently while on the road can be challenging for RVers and travelers, who must ensure an ample supply of food, water, and other necessities such as batteries. Maximizing storage space while selecting compact yet versatile supplies are effective strategies for stockpiling essentials without overstuffing their RV or travel trailer.

Investment in space-saving storage solutions such as collapsible containers, vacuum sealed bags and stackable bins is an efficient way to maximize available storage space and make essential supplies more accessible and easier to organize and access. Furthermore, prioritizing multipurpose items and compact, lightweight products will further minimize space requirements without sacrificing functionality or convenience.

When stockpiling food and water supplies, prioritize non-perishable items with long shelf lives such as canned goods, dry fruits, nuts, grains, dehydrated or freeze-dried meals, dehydrators/freezer dryers as well as portable water purification systems or tablets to ensure access to clean drinking water while traveling - especially camping in remote locations without access to potable sources of drinking water.

Utilizing regular inventorying and rotation can ensure items stay fresh, usable, and within their expiration dates. Consider creating a designated pantry area within your RV to store all stockpiled supplies securely yet accessible during travel.

By adopting creative storage solutions, prioritizing space-saving items and selecting essential supplies with care, RV owners and travelers can efficiently stockpile essential supplies while taking part in extended adventures with confidence and ease.

## Cooking and Food Prep in an RV Kitchen

Preparing meals while traveling presents unique culinary challenges, limited space, resources, and cooking appliances present unique hurdles. But with creative thinking and strategic use of available resources RV owners can enjoy delectable culinary adventures on the road while enjoying tasty meals to bring their culinary imagination alive!

Maximizing RV kitchen counter and storage capacity is crucial to efficient meal preparation and organization. Investing in collapsible or stackable cookware, compact gadgets, and multipurpose appliances will maximize available space while minimizing clutter while cooking.

As part of your RV travel meal planning strategy, prioritize recipes which require minimal prep, cooking time and cleanup efforts. One-pot meals, sheet pan dinners and slow cooker recipes offer convenient solutions that require few dishes while making use of limited kitchen resources and space.

Securing your RV kitchen with essential pantry staples like dried herbs and spices, condiments, shelf-stable sauces, and basic baking ingredients ensures you will always have access to everything needed to create delicious meals on the road. Consider investing in a compact spice rack or organizer to keep everything organized during meal preparation.

Out-of-door cooking options such as portable grills, camp stoves and fire pits allow you to expand your culinary horizons when camping or traveling in an RV. Cooking over an open flame lends meals an exciting new taste while taking full advantage of outdoor adventure travel lifestyle.

RV owners who adopt efficient storage solutions, plan meals carefully, and embrace outdoor cooking options can enjoy delicious cuisine and fulfilling culinary experiences while traveling across America's wide-open spaces.

Navigation and Route Planning for Off-Grid Travel Navigation and route planning skills are indispensable skills for off-grid travelers and RVers who wish to safely explore remote places or navigate unfamiliar terrain safely and efficiently. Because GPS navigation systems and mapping apps may not always be reliable or accessible while traveling off the grid, having alternative tools and strategies in place for navigation becomes even more essential.

Before embarking on off-grid travel, conduct in-depth research. Collect information regarding road conditions, potential hazards and amenities or services along the journey. Use paper maps, topo maps or satellite imagery to familiarize yourself with terrain; map landmarks such as water sources or potential camping spots as part of this process.

**Make the smart decision - invest in a reliable GPS navigation device or download offline maps onto your phone or tablet for seamless navigation even in areas without cell service coverage**, as well as backup devices like handheld GPS units and compasses just in case technology or signal issues arise.

Plan off-grid routes carefully, giving priority to safety and accessibility by selecting routes suited for the size, weight, and off-road capabilities of your vehicle. Be prepared for unexpected road closures, detours, or obstacles by having alternative routes or contingency plans prepared ahead of time.

Maintaining situational awareness while traveling off-grid is critical to adapting to ever-evolving conditions and having an enjoyable journey. By taking proactive steps when planning and navigating off-grid routes, RVers can explore remote destinations with confidence and peace of mind.

## Wilderness Survival Skills for RVers

Knowing basic survival techniques in remote or wilderness locations where amenities, services and assistance might be limited or unavailable is invaluable when traveling off-grid and exploring remote destinations. Acquiring these essential survival techniques will not only increase safety but resilience as well.

RVers need a range of wilderness survival skills to effectively traverse their RV through wilderness terrain:

- Shelter Construction: Learning how to quickly build temporary shelters using natural materials or camping gear can provide protection from harsh elements as well as warmth and safety in emergency situations.
- Fire-Starting Techniques: Mastering fire-starting techniques using various means like friction, flint and steel or flame starters will enable access to warmth, light and cooking capabilities while exploring wilderness environments.

- Water Acquisition and Purification: Knowing how to obtain, collect, and purify water from natural sources like streams, lakes or rivers is crucial for staying hydrated in the wilderness and avoiding dehydration.
- Navigating: RVers who wish to become adept navigators must become experts at map reading, compass use, and orienteering techniques to successfully navigate unfamiliar terrain and find their way home if lost or disoriented. Developing these abilities enables RVers to navigate unfamiliar territory efficiently while finding their way home quickly should they get lost or disoriented.
- First Aid: Arming themselves with basic first aid knowledge and wilderness medicine skills equips RVers to respond appropriately in the case of injuries, illnesses or medical emergencies when travelling off grid.

RVers who acquire and practice wilderness survival skills will strengthen their readiness, resilience, and ability to respond appropriately in emergencies or unexpected challenges while exploring remote or wilderness regions.

## Waste and Sanitation in an RV

Proper waste and sanitation management in an RV is integral for maintaining cleanliness, hygiene, and environmental responsibility when traveling or living on the road. Implementing effective waste practices helps eliminate unpleasant odors while protecting contamination risks as well as meeting all required compliance measures when camping out in remote or environmentally sensitive regions.

## Waste and sanitation management.

- Waste Disposal: Proper disposal of gray water (from sinks, showers, and appliances) and black water (from toilets) is essential to curb pollution and minimize environmental impact. Designated dump stations or sanitary sewer connections offer reliable methods for disposing waste tanks responsibly and disposing sewage responsibly.
- Waste Reduction: Reducing waste generation through conscious consumption, recycling and composting helps save resources while traveling off-grid and reduces frequent trash disposal needs.
- Odor Control: By employing products like tank treatments, deodorizers, and ventilation systems for RV interior odor management, using products to neutralize smells can help create an inviting interior environment in an RV.
- Hygiene practices: Proper personal hygiene habits such as handwashing, dishwashing and surface cleaning help combat germs, bacteria, and pathogens from spreading in an RV environment.
- Environmental Stewardship: Care for our environment by respecting natural environments, adhering to Leave No Trace principles and following camping rules and regulations as part of responsible RVing promotes responsible RVing while safeguarding wilderness areas for future generations.

RVers can reduce their environmental footprint while enjoying an enjoyable travel experience in comfort and sustainability on the road by adopting responsible waste management practices and prioritizing cleanliness and hygiene.

## Security Measures for RVs and Motorhomes

Security measures are key components to protecting RVs and motorhomes against theft, vandalism and unwarranted access when traveling or parking them in unfamiliar or distant places. By employing multiple measures to secure them against intruders while safeguarding valuable assets and belongings.

## RV and motorhome security

Door and Window Locks: Installing high-quality exterior door and window locks is essential to protecting against unauthorized entry into an RV when parking or leaving unattended. Installing quality locks helps safeguard its contents as well.

- Alarm and Security Systems: Equipping an RV with alarm systems, motion sensors and GPS tracking devices provides additional layers of security and alerts occupants to any security breaches or attempted breaches that might take place.
- Exterior Lighting: Installing exterior lights such as motion activated or LED floodlights helps deter intruders while increasing visibility around an RV in low light settings, particularly at night or when visibility may be poor.
- Storage Compartment Locks: Locking exterior storage compartments, hatches and access panels with locks or padlocks will protect valuables stored there from being stolen and will increase protection of stored belongings.
- Steering Wheel Locks and Chocks: Installing steering wheel locks, wheel chocks or tire locks helps immobilize an RV while in parking or storage to prevent unauthorized movement or theft of its contents.

Regular inspection and maintenance of vehicle security components helps maintain their effectiveness and reliability over time, as does practicing situational awareness, parking in well-lit and secure areas and remaining alert for suspicious activity while traveling - these steps reduce the risks of vehicle thefts or break-ins significantly while traveling.

Implementing robust vehicle security measures and adopting proactive security practices are integral ways RVers can safeguard their vehicles, belongings, and personal safety while exploring adventures on the road.

## DIY Project: RV Survival Kit: Preparing Your Mobile Shelter

**Materials:**

- Water filtration system
- Non-perishable food supplies
- First aid kit
- Emergency radio
- Flashlights and batteries
- Multi-tool
- Fire extinguisher.
- Emergency blankets
- Portable toilet

- Rope or paracord
- Duct tape.
- Solar charger
- Portable stove or camping stove.
- Waterproof matches or lighter
- Signal mirror.
- Whistle
- Compass
- Map of the area
- Bug spray
- Personal hygiene items

**Procedure:**

- Start by gathering all the materials listed above and placing them in an easily accessible storage area within your RV. Make sure everything can be quickly reached in an emergency.
- Water is essential to survival; ensure you invest in an effective water filtration system to purify any of the drinking water encountered while traveling and have enough storage containers ready.
- Stock your RV with nonperishable food such as canned goods, dried fruits, nuts, and energy bars to provide at least two weeks' worth of nourishment to each resident of your vehicle.
- Assemble an extensive first aid kit by gathering bandages, gauze pads, antiseptic wipes, pain relievers, tweezers, scissors, and any necessary prescription medicines.
- Purchase an emergency radio that will receive weather alerts and broadcasts even in remote areas and keep spare batteries for all electronic devices you own - like radios.
- Pack several flashlights with extra batteries so that in the event of power outages or nighttime emergencies you have reliable lighting available to you.
- Multi-tools are flexible tools designed for various tasks, from repairs to opening cans.
- Include a fire extinguisher in your RV in case of a fire emergency and make sure everyone on board understands its use.
- Emergency blankets are lightweight and compact blankets designed to offer warmth and shelter against the elements should the need arise for sheltering outdoors.
- Portable toilets provide an effective means of maintaining personal hygiene during emergencies or camping trips in remote areas without access to restroom facilities.
- Rope or paracord can be useful when it comes to securing items, creating shelters, or dealing with emergency situations.
- Duct tape can be an indispensable resource, offering fast fixes or quick solutions in times of emergency.
- An off-grid solar charger can keep your electronic devices charged even in remote places, providing reliable electricity supply to them all at the same time. Select one which fits with the device(s) in question as well as has sufficient charging capacity.

- An outdoor camping stove allows you to prepare food and boil water without even access to traditional kitchen facilities.
- Waterproof matches or lighters are essential tools for starting fires in wet or damp conditions.
- Signal mirror and whistle combinations can help attract attention and signal for assistance if you become lost or distressed.
- Always carry a compass and map of the area when traveling through unfamiliar terrain, particularly if GPS services are unavailable. This may help to ensure safe navigation through unfamiliar territory.
- Bug spray is an invaluable way to defend against insects that pose potential threats and diseases in outdoor settings.
- Make sure that personal hygiene items such as soap, toothpaste and toilet paper are part of your emergency kit to maintain cleanliness and ensure comfort during an unexpected situation.

# BOOK 13: PREPPER'S COMMUNITY: SHARING PREPAREDNESS

Community resilience refers to a community's ability to prepare, respond to and recover from different types of challenges, crises or disasters while remaining cohesive, well-functioning units. Understanding community resilience requires acknowledging how interdependent all members, families, organizations, and institutions in a community are and the collective impact they can have in mitigating shared vulnerabilities while improving preparedness. Building community resilience entails cultivating strong social networks, trust, communication, collaboration, and resourcefulness among community members to effectively anticipate, mitigate, manage, and mitigate risks and crises. Through investing in community resilience-building efforts like education, training, infrastructure development and social support systems, communities can increase their ability to adapt, recover and thrive despite adverse situations and uncertainty.

## Analyzing Community Strengths and Weaknesses

Assessing community strengths and weaknesses is integral for developing targeted resilience-building strategies and interventions tailored to address specific challenges that face any given community. Conducting comprehensive community assessments involves gathering and analyzing information regarding several aspects, including demographics, socio-economic status, infrastructure resources vulnerabilities as well as experience dealing with disasters or emergencies. Engaging community members, stakeholders, and experts through surveys, interviews, focus groups and community forums is the best way to identify strengths, assets, capabilities, and areas for improvement in any community. Work closely with local agencies, organizations, and institutions to leverage existing resources, expertise, initiatives, and initiatives while effectively addressing gaps and vulnerabilities in communities. Recognizing and capitalizing upon community strengths while simultaneously tackling weaknesses can strengthen resilience by giving communities more strength to withstand and recover from adverse events.

## Emergency Response Teams

Forming emergency response teams within communities is a proactive measure designed to increase preparedness, coordination, and response capabilities during times of crises or disaster. Establishing dedicated teams of trained volunteers or professionals enables communities to mobilize resources, expertise, and manpower quickly and effectively when responding to emergency situations. Form emergency response teams composed of people from diverse backgrounds and expertise - medical professionals, first responders, community leaders and volunteers who possess training in disaster response, search and rescue, medical care delivery, communication logistics coordination. Provide ongoing training, drills, and exercises to prepare team members to respond swiftly and collaborate successfully with other agencies during emergencies. Foster strong partnerships and collaboration among emergency response teams, local authorities, government agencies, and community organizations to promote coordinated emergency response efforts. By developing comprehensive emergency response teams within communities can improve resilience and capacity during emergencies by protecting lives, property, livelihoods and livelihoods from threats or emergencies.

## Establishing Neighborhood Watch Programs

Establishing neighborhood watch programs can be an excellent way of strengthening community safety, security, and resilience by mobilizing residents in proactive measures against crime, vandalism, or other safety risks in their neighborhood. Neighborhood watch programs involve residents joining together to monitor, report, and address suspicious activities in their community while upholding an environment of cooperation among members and mutual support among its residents. Establish neighborhood watch committees or groups consisting of volunteers representing diverse backgrounds and demographics to engage the whole community in safety and security initiatives. Provide training, resources, and support to neighborhood watch participants regarding crime-prevention strategies, emergency preparedness procedures, communication protocols and working in cooperation with law enforcement and local authorities. Communication among neighborhood watch participants should take place regularly through meetings, newsletters, social media posts and community events to strengthen community bonds and resilience. By creating neighborhood watch programs within communities can empower residents to play an active role in strengthening safety, security, and resilience at the grassroots level.

## Preparedness Workshops and Training Hosted

Hosting emergency preparedness workshops and training sessions is an effective strategy to strengthen community resilience by arming residents with knowledge, skills, and resources needed to respond quickly to emergencies or disasters and recover quickly afterwards. Arrange workshops and training sessions on various subjects related to emergency preparedness, disaster response, risk mitigation, first aid/CPR training/fire safety procedures/evacuation protocols/communication strategies/community resilience-building. **Cooperate with local agencies, organizations, and experts to devise and deliver tailored training programs** tailored specifically for their community's unique needs, priorities, and vulnerabilities.

Introduce workshops and training sessions through various channels - in-person meetings, webinars, community events - to reach as wide an audience possible and accommodate individuals' varying learning preferences and schedules. Assemble hands-on training, simulations, drills, and exercises designed to engage participants actively while reinforcing key concepts and skills. Engaging all levels of community life--residents, businesses, schools, faith-based organizations, and community groups alike--in creating a culture of preparedness and resilience can foster collective strength. Hosting preparedness workshops or training sessions gives residents access to tools they can use effectively respond to emergencies while building resilience collectively.

## Develop Communications Lists and Networks

Establishing communication lists and networks are vital in providing communities with effective information sharing, coordination and cooperation during emergencies and disasters. Build comprehensive contact lists containing important stakeholders, community leaders, emergency responders, local authorities, government agencies, medical facilities, schools, businesses, and community organizations. Establish communication networks and channels - such as phone trees, email lists, social media groups and community forums - that enable quick dissemination of updates, alerts, and instructions among community members quickly and efficiently. Staying prepared during emergencies requires updating contact lists regularly, verifying contact info and testing communication channels to ensure maximum reliability and accessibility during times of distress. Strengthen relationships and cultivate trust between community members, stakeholders, and partners to foster open communication, cooperation, and assistance when times get difficult. Through communication lists and networks, communities can increase their capacity to disseminate crucial information during emergencies or disasters as well as mobilize resources more efficiently when the time comes.

## Establishing Relationships with Local Agencies

Building relationships with local authorities is integral to fostering collaboration, cooperation and trust between communities and government agencies in emergency preparedness, response, and recovery efforts. Establish channels of communication and collaboration between emergency management agencies, law enforcement, fire departments, public health departments and municipal governments in your region. Engage with local officials, policymakers, and decision-makers to advocate for community needs, priorities and concerns related to emergency preparedness and resilience-building. Take part in community meetings, advisory boards, task forces and committees to provide input, guidance and direction regarding emergency planning policies and initiatives. Collaborate with local authorities on joint training exercises, drills, and tabletop simulations to enhance coordination and interoperability between community-led response efforts and government response efforts. By forging relationships with their local authorities, communities can tap their expertise, resources, and support to strengthen resilience against emergency and disaster events and respond appropriately.

## Organization of Aid Networks and Mutual Aid Organizing

Establishing aid networks and mutual aid initiatives are effective means of strengthening community resilience and solidarity by offering mutual assistance among community members during emergencies or disasters. Establish mutual aid networks within their community comprising residents, businesses, organizations, and institutions to pool their resources, skills and expertise for mutual advantage and collective action.

Protocols, guidelines, and agreements for mutual aid exchanges such as resource sharing, volunteer deployments, sheltering needs or supporting recovery efforts need to be in place for mutual aid exchanges to function successfully. Encourage community members to get involved with mutual aid initiatives by offering their time, talents, resources, and support in times of need. Create an environment of reciprocity, empathy, and solidarity within your community to foster it as an essential value and practice. Through aid networks or initiatives organized specifically around mutual aid initiatives communities can strengthen social cohesion while building trusting relationships that strengthen support during challenging periods and help each other out more easily.

## Conduct Risk Evaluations

Conducting risk analyses is an integral component of improving resilience and preparedness against emergencies or disasters in communities. They enable identification, understanding, mitigation and reduction of hazards, vulnerabilities, risks as well as hazards within them to foster enhanced community preparedness for emergency and disaster responses. Engaging community members, stakeholders, and experts to conduct comprehensive risk analyses which encompass various hazards like natural disasters, technological hazards, public health emergencies and human-caused events can help conduct thorough risk analyses that provide accurate assessments. Gather historical events, exposure, susceptibility, and consequences information regarding potential hazards to aid risk assessment processes.

Analyze the likelihood and impact of various hazards on community assets, infrastructure, population groups and critical facilities. Recognizing and mitigating high-risk areas within your community are necessary steps to reduce risks and build resilience. Develop risk mitigation strategies, action plans and prioritized interventions based on risk analysis findings to strengthen community resilience and preparedness. Conducting risk evaluations allows communities to identify vulnerabilities, gain a better understanding of potential hazards, prioritize actions to reduce risks effectively while strengthening resilience effectively.

## Establishing Emergency Shelters and Spaces

Establishing emergency shelters and safe spaces are crucial in providing temporary refuge, support, and aid to individuals affected by emergencies and disasters. Locate suitable locations, resources and facilities within your community that could serve as emergency shelters - schools, community centers, churches or public buildings may serve this function effectively. Prepare these spaces ahead of time with essential supplies, equipment, and amenities such as cost, blankets, food, water, hygiene facilities, medical supplies, and security measures. Plan, protocols and procedures must be established to activate and manage emergency shelters effectively - registration, intake, security monitoring, supervision oversight as well as coordination between local authorities and support agencies must all take place simultaneously. Staff and volunteers should receive training on providing compassionate care, assistance, and support to shelter residents while respecting their dignity, privacy, and rights. Emergency shelters must ensure accessibility, inclusivity, and accommodation of individuals with disabilities, older individuals, families with young children and other vulnerable groups in emergency shelters and spaces. By doing this, communities can quickly provide aid during an emergency and disaster to ensure resilience and recovery is promoted quickly.

## Promoting Diversity and Inclusion in Preparedness Efforts

Promoting diversity and inclusion within preparedness efforts is paramount to providing fair access to resources, information, and support for all members of a community, regardless of race, ethnicity, nationality, religion, gender, age, ability, sexual orientation, or socioeconomic status. Engaging diverse community groups, organizations, leaders and stakeholders in emergency preparedness and resilience-building is critical. Listening closely to their needs, concerns, perspectives, and priorities regarding emergency preparedness is imperative to creating resilience-building plans that stand a chance in disaster situations.

Create inclusive outreach, communication and education strategies that engage all segments of the community - particularly marginalized and underrepresented populations. Provide culturally and linguistically relevant resources, materials, and information regarding emergency preparedness, response and recovery in multiple languages and formats. Foster partnerships and collaborations between community-based organizations, faith-based groups, advocacy organizations, and social service agencies that serve diverse populations to bolster preparedness efforts and address any special needs and vulnerabilities of marginalized groups. By prioritizing diversity and inclusion within preparedness initiatives, communities can strengthen social cohesion, build trust among all residents involved, and empower all to participate and contribute to resilience building efforts.

## Strengthening Social Cohesion Through Shared Experiences

Strengthening social cohesion through shared experiences is integral to building trust, solidarity, and resilience among members of a community - particularly during times of difficulty or crisis. Create opportunities for community members to unite, connect, and form lasting bonds through shared experiences, activities, events, or initiatives. Establish community gatherings, festivals, fairs, workshops, and volunteer projects which foster social interactions and collaboration within your area of residence while strengthening mutual support relationships between members. Encourage storytelling, dialogue and reflection about shared experiences, challenges, and triumphs related to emergency preparedness, response, and recovery.

Fostering an atmosphere of belonging, empathy, and interconnectedness among community members by celebrating diversity, honoring traditions, and appreciating individual contributions is crucial in creating an atmosphere that values everyone's contributions and perspectives. Resilient communities need platforms and spaces where residents can share knowledge, skills, resources, and solutions for building resilience collectively. By strengthening social cohesion through shared experiences, communities can increase their capacity to support each other while building trust between members that enables them to tackle challenges together more successfully.

## Implementing Sustainable Practices for Resilience

Implementing sustainable practices for resilience involves adopting eco-friendly, resource-saving, climate-resilient initiatives that promote long-term well-being and resiliency within communities. Integrating principles of sustainability, conservation, and resilience into community planning projects to reduce vulnerability against environmental hazards and climate change impacts. Implement energy efficiency measures such as renewable energy sources, water conservation practices and waste reduction to minimize ecological footprints and build resilience against resource scarcity and environmental degradation. Enhance sustainable agriculture, food production and land management practices that support local food security, biodiversity conservation and ecosystem health.

Foster community resilience through eco-friendly economic development, job creation, and entrepreneurial initiatives that focus on social equity, economic diversity and building local resilience--such as sustainable economic development projects or employment programs that prioritize these qualities. By adopting sustainable resilience-enhancing practices communities can enhance adaptive capacity while decreasing vulnerabilities while simultaneously creating equitable environments that support current as well as future generations' wellbeing and wellbeing.

## Supporting Vulnerable Populations in the Community

Supporting vulnerable communities within communities is integral for providing equal access to resources, services and assistance during emergencies or disasters. Identification and prioritization of needs, concerns and vulnerabilities associated with vulnerable populations - elderly adults, people with disabilities, low-income families, immigrants/refugees/homeless individuals etc. - in terms of needs/concerns/vulnerabilities to provide needed interventions that help close social, economic or health disparities and inequities. Utilize targeted outreach, education and assistance programs designed to meet the specific needs and challenges of vulnerable populations in emergency preparedness, response, and recovery efforts.

As part of your efforts to aid vulnerable populations, seek partnerships and collaborations with social service agencies, nonprofit organizations, faith-based groups, and community-based groups specializing in aiding. Doing this will streamline support and assist efforts. Provide accessible and culturally sensitive resources, information, and services that equip vulnerable populations to prepare and respond effectively for emergencies in their community. By helping vulnerable groups prepare, respond, and recover more easily during emergencies or disasters. Support vulnerable residents as a means of building social equity, inclusion, resilience, and strengthening the strength and potential for all residents in recovery efforts from emergencies and disasters.

## Fostering a Culture of Preparedness Through Proactive Engagement

Proactive engagement from all community members is necessary in creating an atmosphere of preparedness within an area. Encourage community members to take ownership over their preparedness efforts by offering education, resources, and support services necessary for proactive planning and action. Create workshops, training sessions and community events that foster preparedness and resilience-building activities. Empower individuals and families to develop customized emergency plans, assemble disaster kits and stay abreast of potential risks or hazards. Bring about collective responsibility and mutual aid among neighbors by encouraging collaboration in times of need and proactive engagement in preparedness efforts, which will build resilience and increase capacity to withstand and recover from emergencies or disasters.

## Leveraging Community Resources for Disaster Recover

Utilizing local assets, capacities, and networks for disaster recovery within communities requires mobilizing assets that enhance response and recovery efforts as part of an overall response and recovery effort. Recognize and map existing resources within your community - including facilities, equipment, expertise, volunteers, and support services that could be mobilized during disaster response and recovery efforts. Establish partnerships and collaborate with local businesses, organizations, institutions, and government agencies to utilize their resources and capabilities for mutual aid and assistance. Create resource sharing agreements, mutual aid arrangements and memoranda of understanding to facilitate coordination and cooperation among stakeholders during emergencies or disasters. Leveraging resources within communities allows communities to increase resilience by meeting immediate needs as well as long-term challenges of recovery more efficiently.

## Reconciling Resilience Efforts: Recognizing and Praising Resilience Strategies

Recognizing and celebrating resilience efforts within your community is vital to engender pride, solidarity, and motivation among residents and stakeholders alike. Recognizing and honoring individuals, organizations, or groups who have demonstrated exceptional resilience, leadership, or contributions towards community preparedness or recovery efforts should be recognized with awards or accolades. Host recognition events, awards ceremonies, and appreciation activities to recognize resilience achievements in your community as well as showcase success stories. Showcase inspiring stories of resilience and perseverance to motivate others and cultivate a culture of resilience in communities. By acknowledging and celebrating resilience efforts, organizations can strengthen morale, cohesion, and confidence that their members can overcome challenges successfully and thrive despite hardship.

## Advocating for Policy Changes that Support Community Resilience

Advocating for policy changes that promote community resilience involves engaging policymakers, legislators, and government agencies in efforts to enact laws or policies which strengthen preparedness, mitigation, and recovery efforts. Advocate for policies and initiatives that prioritize investment in infrastructure, resources, and programs designed to increase community resilience while simultaneously addressing vulnerabilities and inequalities in society. Support measures which promote climate adaptation, disaster risk reduction and sustainable development to build resilience against future hazards and challenges. Collaborate with advocacy groups, coalitions, and stakeholders to amplify voices and advocate for policy changes on local, state, and national levels. By doing this, communities can foster an environment conducive to resilience building efforts while building sustainable, equitable and resilient societies.

## Integrating Technology and Alerts for Emergency Communication

Integrating technology and alerts for emergency communication is crucial in providing timely, accurate, and efficient dissemination of vital information during times of emergencies and disasters. Employ emergency alert systems like reverse 911, text messaging, email notifications, social media notifications and mobile apps to quickly inform residents with updates, warnings, or instructions they require. Utilize social media, websites, and digital communication channels to share real-time information, updates, and resources with the community in real-time. Use GIS mapping, satellite imagery and other technology tools such as Gmail alerts to monitor hazards and track impacts before organizing response efforts in response to emergencies or disasters. By incorporating technology and alerts for emergency communication purposes, communities can improve situational awareness while strengthening public safety while expediting response/recovery initiatives.

## Continuous Evaluation and Enhancing Resilience Plans

Evaluating and enhancing resilience plans regularly is crucial to keep community preparedness efforts up to date with emerging risks and threats. Regular reviews should be performed to evaluate and assess existing resilience plans, policies, and procedures to identify strengths, weaknesses, gaps, and opportunities for improvements. Engaging community members, stakeholders, and partners through surveys, focus groups, or post-action reviews in gathering feedback will assist evaluation efforts. Update resilience plans, strategies, and protocols based on lessons learned, best practices, and emerging trends in emergency management and resilience-building. Conduct training exercises, drills, and tests to test, validate, build capacity for, and enhance readiness among stakeholders. By continuously assessing and updating resilience plans communities can adapt more easily to changing conditions; address emerging threats more swiftly while strengthening their capacity to withstand emergencies or disasters efficiently.

## DIY Project: Community Emergency Preparedness Plan: Strengthening Bonds in Crisis

**Materials:**

- Paper and pens
- Map of the community
- Contact information for local emergency services.
- Emergency supply checklist
- First aid supplies
- Communication devices (phones, walkie-talkies, etc.)
- Flashlights and batteries
- Emergency food and water supplies
- Blankets or sleeping bags.
- Basic tools (hammer, nails, screwdriver, etc.)
- Fire extinguishers.

- Whistles or signaling devices.
- Megaphone or loudspeaker.
- Reflective vests or clothing
- Tarps or tents for shelter
- Portable generators
- Portable cooking equipment
- Hygiene supplies (soap, hand sanitizer, etc.)
- Trash bags.
- Compass and maps of surrounding areas

**Procedure:**

- Begin by gathering community members together for an initial meeting about emergency preparedness and collective action plans, encouraging everyone present to contribute ideas and resources from within their respective fields of expertise.
- Create a list of emergency supplies and resources needed by the community in the event of a disaster, including food, water, medical supplies, communication devices, tools, and shelter materials.
- Map out your community and identify key locations such as evacuation routes, emergency shelters, **medical facilities and gathering points. Mark these on a map and distribute copies to all community members.**
- Make a list of emergency contacts such as police, fire, ambulance services and utility companies and ensure everyone knows how to access this information quickly in an emergency.
- Develop a communication plan for the community that includes leaders or coordinators charged with disseminating information and organizing response efforts, along with alternate channels in case traditional ones cannot.
- Plan regular training and drill sessions to familiarize community members with emergency procedures and protocols, including first aid, fire safety measures, search-and-rescue techniques, and communication strategies. Practice key competencies such as first aid, fire safety practices and search and rescue techniques as well as communication techniques.
- Establish a system to identify vulnerable members of your community, such as senior citizens, those with disabilities and those who require special medical assistance, to develop plans that ensure their wellbeing during emergencies.
- Encourage community members to develop emergency kits for themselves and offer guidance as to which items they should include in their kit. Aid if anyone requires help creating their kit.
- Implement a buddy system or neighborhood watch program to strengthen community bonds and support networks, encouraging residents to check in during emergencies and offer help if needed.
- Partnerships between local organizations, businesses, and government agencies can strengthen community resilience while opening the way to additional resources and support.

Regularly review and revise the community emergency preparedness plan considering changing community conditions or emerging risks or lessons learned through experience. Encourage ongoing participation by all members of your community.

# BOOK 14: ENERGY CREATION IN SURVIVAL SCENARIOS

Energy is crucial to our daily survival as it powers essential functions like heating, lighting, cooking, communication, and transportation. Access to reliable sources can make all the difference in emergency or survival scenarios; accessing reliable energy sources could mean the difference between comfort and discomfort, safety from danger or even life and death. Recognizing energy's role requires acknowledging its key function of maintaining basic human needs while powering critical equipment as well as supporting essential tasks for resilience and survival.

## Assess Energy Needs and Scenarios

Assessing energy scenarios and needs involves analyzing a survival situation to ascertain its specific requirements and challenges to select suitable energy solutions. Consider factors like location, climate, duration of emergency events, available resources and energy requirements of essential activities that require power supply - heating, lighting, cooking, communication medical equipment as well as powering electronic devices among them are critical energy needs that need consideration in this assessment process. Finally evaluate existing energy sources infrastructure equipment along with any risks or vulnerabilities which might hinder its distribution.

## Utilizing Renewable Energy Sources

**Renewable energy sources like solar, wind and hydro power offer sustainable and scalable solutions for survival situations. Solar harnessing involves harnessing sun energy with photovoltaic panels for electricity production; wind power uses turbines to convert wind energy to electrical energy production; hydro power uses flowing water as its power source via turbines or waterwheels to produce electrical outputs - each has their own distinct set of advantages for survival such as reliability, scalability independence from fuel supplies as well as environmental sustainability.**

## Portable Energy Solutions

Portable energy solutions such as generators and power banks are crucial tools in emergency scenarios, especially remote or off-grid areas. Generators use fuel such as gasoline, diesel, propane, or natural gas to supply electricity on demand for various uses while power banks store energy stored in rechargeable batteries that allow users to charge electronic devices such as phones, radios, flashlights, and medical equipment - offering flexibility, mobility, and convenience during critical moments of survival.

## DIY Energy Generation

DIY energy generation involves building and using homemade wind turbines and solar panels for survival situations, using simple materials and designs to harness wind energy to produce electrical power; similarly homemade solar panels may utilize photovoltaic cells mounted onto glass frames that capture solar radiation to convert into energy for use as electrical power production. DIY energy generation provides cost-effective yet customizable solutions for off-grid living scenarios while decreasing reliance on centralized power grids and more efficiently harness renewable sources of power generation.

Understanding the significance of energy for survival involves evaluating energy scenarios and needs, tapping renewable sources for power production, exploring portable solutions for portable energy consumption needs and considering DIY energy generation options. By adopting an all-inclusive strategy to energy planning and preparation, individuals and communities alike can increase resilience, self-sufficiency, and survival capabilities under demanding circumstances.

## Batteries Vs Capacitors

When it comes to energy storage options for various purposes, batteries and capacitors are two popular choices that each come with their own set of advantages and drawbacks.

- Batteries are widely used as energy storage units due to their chemical makeup and ability to hold larger amounts than capacitors compared to portable electronics and vehicles; in renewable energy systems too as a power source that delivers sustained power for extended periods.
- There are various kinds of batteries on the market today, such as lead-acid, lithium-ion, and nickel metal hydride cells - each offering different energy densities, lifespan, and charging/discharging abilities.
- Batteries provide higher energy density and longer storage periods compared to capacitors; however, their weight, bulkiness, and charging rate make them heavier, bulkier, and slower compared to charging times with capacitors. They have a limited lifespan and may need regular maintenance or replacement over time.
- Capacitors store energy in an electric field and offer rapid charging/discharging capability, making them suitable for applications requiring quick bursts of power such as cameras, flashlights, and power tools.
- Capacitors offer lower energy density compared to batteries, meaning that they store less power per volume or weight; however, their reduced internal resistance allows more effective energy transfer and quicker response times.
- One advantage of capacitors over batteries is their longer lifespan and higher cycle life; unlike batteries they do not deteriorate with repeated charging/discharging cycles like batteries do and therefore are eco-friendlier and more recyclable than certain battery types.

## Improvising Emergency Energy Sources in Critical Situations

In emergencies where conventional energy supplies may be limited or nonexistent, improvisation becomes crucial in providing access to essential power supplies. There are various techniques which can be implemented in emergency energy production:

- Solar Energy: Solar panels harness sunlight energy and convert it to electricity for emergency power needs in emergencies and on an everyday basis, offering renewable and eco-friendly sources of power that don't depend on fossil fuels like power from traditional sources like utility companies or generators. Portable chargers or generators may also be available as additional backup sources of electricity needed by small electronic devices or essential appliances that cannot be charged via traditional means alone.
- Wind Power: Wind turbines can be an efficient means of harnessing wind energy for power production in areas with consistent winds patterns, serving as reliable sources of electricity generation. Portable wind generators or DIY wind generators may also be created using simple materials and components for emergency backup power supplies.

- Human Power: Hand-crank or pedal generators use manual labor to generate electricity that can be used for charging batteries, powering small electronics devices, or operating essential appliances in emergency situations. These portable machines may prove beneficial.
- Generator Fuel: Portable generators using gasoline, diesel, propane, or natural gas provide essential backup power when emergencies strike. Therefore, it's imperative that an adequate supply of fuel be stored away and regularly maintained to ensure reliable functionality when required.
- Pre-charged Battery Banks or Power Banks: Backup power banks offer an effective means for charging electronic devices or operating small appliances during power outages or emergencies, offering vital backup power sources.
- Employing these improvisational methods, individuals can utilize essential energy sources during critical moments when conventional power sources may become disrupted or unavailable, providing access to essential supplies of vital energy and maintaining functionality in critical scenarios where conventional sources might become disrupted or unavailable.
- Optimizing Energy-Efficiency Devices and Appliances for Energy Conservation In off-grid or emergency scenarios with limited energy resources, prioritizing energy-efficient devices and appliances is key to conserving power and stretching available resources further. Multiple strategies exist that can maximize energy efficiency while decreasing consumption:
- **LED Lighting: Energy-saving** LED lightbulbs consume much less power than incandescent or fluorescent bulbs, helping you lower energy consumption for lighting purposes and extend battery lifespan in emergency or off-grid scenarios. Switching over can save money as well as battery power.
- Energy Efficient Appliances: When selecting energy-efficient appliances with Energy Star ratings, choosing those featuring variable speed motors, insulation, and programmable settings to reduce energy use during everyday household activities like cooking, refrigeration and water heating will significantly cut energy consumption costs and consumption rates.
- Insulation and Weatherization: Insulating and weatherizing a home or living space will help create more comfortable indoor temperatures while decreasing heating or cooling energy needs. Seal drafts, insulate windows and doors, use thermal curtains or shades, seal drafts in drafty places such as basements or attics, use draft dodgers in drafty corners to block drafty doors in winter weather to minimize heat loss as well as reduce cooling loads during hotter spells.
- Smart Power Management: Use smart power strips, timers, and programmable thermostats to efficiently monitor energy use for electronic devices, appliances, and heating/cooling systems. Time your operations around peak solar or wind production periods to optimize renewable energy utilization and ensure maximum renewable resource exploitation.
- Energy Efficient Cooking Methods: Opting for energy-saving cooking methods like pressure, induction or solar cooking is one way to cut energy usage when creating meals. Such techniques reduce cooking times, wasteful energy use and fuel expenses when compared to conventional appliances.

By prioritizing energy-saving practices and devices and appliances, individuals can reduce energy usage, decrease reliance on external energy sources and enhance sustainability in off-grid or emergency scenarios.

## Prioritizing Energy Usage to Achieve Survival in Critical Scenarios

Energy usage plays an essential role in meeting basic human needs during times of limited access or uncertainty; prioritization allows individuals to better allocate available resources and increase chances of survival during emergency scenarios by recognizing critical energy requirements and prioritizing accordingly.

Prioritize energy use for water purification and sanitation to provide access to safe drinking water and avoid waterborne illness, making energy-saving methods such as solar water disinfection or portable filters an efficient means of purifying without depending on conventional energy sources such as fossil fuels.

- Heating and Shelter: Prioritize energy-saving heating methods such as insulated shelters, passive solar heating, or portable heaters to protect yourself against extreme temperatures or weather conditions while simultaneously providing warmth and comfort. Energy-efficient methods may include using energy-efficient buildings like Passive solar heating panels or energy efficient portable devices which save both money and resources while offering much needed warmth and coziness for essential warmth and relaxation.
- Communication and Information: Set energy-use priorities to stay abreast of emergency alerts, weather forecasts, evacuation notices and alerts related to safety. Use efficient communications devices like handheld radios, satellite phones or solar-powered chargers that conserve battery power while maintaining communication capabilities.
- Food Preservation and Storage: Prioritize energy usage to preserve perishable foods without depending on grid power; energy-efficient refrigeration methods like insulated coolers, root cellars or solar-powered fridges should be utilized as effective ways to extend their shelf lives while saving grid energy usage.
- Medical Care and First Aid: Prioritize energy usage for medical care and first aid purposes to address injuries, illnesses, or medical emergencies effectively. Provide access to essential devices, medications and supplies while prioritizing energy allocation to power medical equipment, lighting devices and communication devices.

Prioritizing energy usage to meet essential survival needs can help conserve resources, optimize allocation, and strengthen resilience during emergency scenarios where power may be limited or unreliable.

## DIY Project: Portable Solar Generator: A Versatile Energy Solution in Survival Scenarios

**Materials:**

- Solar panels
- Deep-cycle batteries
- Charge controller.
- Power inverter
- Portable generator enclosure or box
- DC disconnect switch.
- Battery cables
- Wire connectors
- Circuit breakers

- Mounting hardware (brackets, bolts, etc.)
- Solar panel cables
- Junction box
- Waterproof sealant
- Insulation material (foam, rubber, etc.)
- Electrical tape

**Procedure:**

- Select an ideal solar panel size and quantity according to your energy needs and availability of sunlight in your region. High-efficiency panels designed with durable designs suitable for outdoor usage should be prioritized when making this selection.
- Mount the solar panels securely to an area with maximum exposure to sunlight, such as a rooftop or open field, using sturdy mounting hardware for optimal sunlight absorption. Orient these solar panels at optimal angles so as not to prevent optimal sunlight absorption by anchoring and tilting them in appropriate ways.
- Connect the solar panels to a charge controller using their respective cables; the charge controller regulates electricity flow to prevent overcharging of batteries.
- **Install the deep-cycle batteries inside of the portable generator enclosure or box to protect them from potential damage and temperature variations. Make sure they are securely mounted and insulated against damage and temperature swings.**
- Connect the batteries to a charge controller using battery cables with wire connectors ensuring secure and reliable connections between components.
- Integrate a power inverter in the generator enclosure to convert DC power stored in batteries into AC current suitable for running devices and appliances.
- Connect the power inverter to the batteries using appropriate wiring and circuit breakers to protect the system against overloads or short circuits.
- Install a DC disconnect switch between your batteries and inverter for safety measures; this switch enables you to quickly isolate your system in case of maintenance work or emergencies.
- Install a junction box within the generator enclosure to organize and secure wiring connections, sealing it off with waterproof sealant to avoid moisture damage to prevent shorts in connections.

Make sure that everything works smoothly by testing out your portable solar generator regularly, tracking performance closely and adjusting as required to maximize efficiency.

# BOOK 15: SURVIVAL SKILLS IN THE WILD

In wilderness survival situations, building shelter is an indispensable skill essential to survival. Tarps provide lightweight yet flexible materials which can easily fit in survival kits or backpacks and should be considered when creating emergency tarp shelters: When creating one with these materials consider these steps when setting it up:

- Selected an Appropriate Location: Select an area free from flood zones, falling debris or animal habitats as the desired shelter location.
- Gather Materials: Collect additional material like branches, logs, rocks and foliage as support structures and insulation layers for your shelter.
- Set Up the Tarp: Stretch out a tarp between trees or secure it to makeshift poles using ropes or cordage, positioning it to provide sufficient wind, rain, and sunlight protection. Ensure it provides optimal positioning to provide adequate shelter from winds, rain, and sunrays.
- Secure Edges: Anchor the edges of the tarp using rocks, logs, or stakes to keep it from shifting or collapsing during strong winds or severe weather conditions. This will prevent it from shifting or collapsing under its own weight if its anchor points shift during windy or inclement weather events.
- Insulate and waterproof your shelter: Layer natural materials like leaves, pine needles or branches over the tarp for additional insulation and waterproofing protection for shelter use.
- Add Entry and Ventilation: Be sure that there are sufficient openings or flaps in the shelter for entry and ventilation to provide enough airflow while offering adequate protection from the elements.
- Perform Stability Tests: Conduct stability and durability tests on the shelter by applying pressure, making sure it can withstand moderate wind gusts and rainfall conditions.
- By making use of natural materials such as tarps and canvas for emergency shelter building, wilderness survivalists can quickly create temporary refuge from harsh outdoor environments.
- Finding and Purifying Water: Springs, Streams and Sources In wilderness survival situations, finding and purifying water are both essential in terms of maintaining proper hydration levels to stave off dehydration. Springs, streams, and natural sources offer potential sources for drinking water but must first be cleaned of contaminants or pathogens before consumption - here's how you can find and purify wild sources:
- Find Water Sources: Look out for signs of water, such as vegetation, animal tracks or topographical features that point towards springs, streams, or water sources such as springs. Listen out for the sound of running water which could indicate nearby creeks or rivers.
- Collect Water: When collecting water from natural sources like springs, streams, or pools it's best to use containers such as bottles of water or makeshift vessels to collect it - avoid stagnant areas which could contain bacteria and parasites!
- Filter Water: Use a cloth, bandana, or DIY filter to extract larger particles and debris from collected water to decrease turbidity and improve clarity of collected waters.
- Purify Water: Clean collected water using methods such as boiling, chemical treatment or filtration to purify it before consumption. Boiling for at least one-minute kills most harmful organisms like bacteria and parasites to ensure safe consumption; alternatively portable filters or tablets may help remove contaminants or pathogens from sources.

- Store Purified Water safely: When transporting purified water for storage or transport purposes, transfer it to clean, sealed containers with their source indicated on them - label them accordingly to track usage and ensure freshness!
- Conserve Water: Employ various water conservation techniques like rationing and sparing use to conserve limited supplies in emergency survival scenarios. Recycling used water provides another use.

By finding, gathering, and purifying water from natural sources in their wilderness environments, wilderness survivalists can ensure they have safe drinking, cooking, and hydration water available for drinking, cooking and use in other ways in the wilderness.

## Fire Starting Techniques: Flint Steel and Bow Plow

Firing starting techniques are integral wilderness survival skills essential for creating warmth, light, and signaling capabilities outside. Flint and steel and bow drill methods have long been utilized as effective approaches for starting fires without modern tools or matches:

- Flint and Steel: To start a fire using flint and steel, gather dry tinder such as grass clippings, leaves, or cotton balls as kindling material. Place the steel against the flint at an acute angle until sparks fly onto your kindling material - continue striking until this spark ignites it completely, then gently blow on any remaining embers to fuel and sustain your flame and form an active fire.
- **Bow Drill: To use this technique of creating friction and spark an ember with ease, combine a bow,** spindle, fireboard, socket and socket holder together using this unique four-step setup: carve out a notch in your fireboard for placing tinder beneath before securely fastening a spindle into your bowstring and spinning rapidly against it while using your bowstring as your leverage source until friction forms into heat; transfer this newly created ember into your tinder bundle before blowing gently on it to spark it off!

Develop the skills to start fires effectively during wilderness survival situations by practicing them now! Gain proficiency and gain confidence when using such methods during emergencies.

## Navigation Skills Are Vital to Survival

Navigation is essential to wilderness survival, helping individuals orient themselves and navigate terrain in search of safety. Different techniques like GPS, map compass navigation and celestial navigation may all be utilized when traversing unfamiliar territory:

- GPS (Global Positioning System): GPS devices use satellite signals to determine your precise location and provide real-time navigation and mapping features. Carry one of these handheld GPS devices or smartphones equipped with this capability as they will help track your location, plan routes, and navigate unfamiliar terrain more effectively than ever.
- Map and Compass Navigation: Map and compass navigation involves using topographic maps and compasses to establish direction, plot routes and traverse terrain. First orient the map according to your surrounding landscape before using a compass to determine your direction of travel by following landmarks, trails, or terrain contours to reach your destination.
- Celestial Navigation: Celestial navigation involves using celestial bodies like the sun, moon, and stars as navigation cues to determine direction, time, and location estimates. Learn to identify these

celestial bodies against the horizon to use celestial cues effectively when traditional tools may not be readily available or reliable enough.

By mastering navigation techniques such as GPS, map and compass navigation and celestial navigation, wilderness survivalists can safely maneuver terrain, find their way home safely, and overcome challenges encountered outdoors.

## Wilderness First Aid: Treating Injuries, Ailments

Knowing wilderness first aid skills is vital in providing immediate medical assistance in remote or wilderness environments where access to professional medical aid may be limited. Master the following wilderness first aid techniques to effectively treat injuries and ailments out there:

- Assessment and Prioritization: Assess and prioritize treatment based on severity and urgency. Conduct a primary survey to detect life-threatening situations like airway obstruction, breathing issues, severe bleeding or shock and provide immediate interventions, as necessary.
- Wound Care: For optimal healing of wounds and injuries, clean and dress them using sterile gauze, bandages, and antiseptic solutions such as gauze with antibiotic properties as a first line defense against infection and to facilitate speedier healing processes. Pressure can also be applied to reduce bleeding. Elevating injured limbs helps minimize swelling and reduce pain relief.
- Splinting and Immobilization: Proper immobilization for fractures, sprains and dislocations using homemade or found objects like branches or trekking poles will prevent further injury to injured limbs and further prevent their breakdown.
- Heat and Cold Injuries: For heat exhaustion or heatstroke-related illnesses, treatment involves cooling the body down with fluids while seeking shade or shelter; cold injuries (hypothermia or frostbite) require warming them through insulation measures and providing warm fluids as treatment options.
- Environment Emergencies: To respond effectively to environmental emergencies like dehydration, altitude sickness or allergic reactions it is crucial that proper measures such as fluid administration, rest and medications or interventions be put in place as soon as possible.
- Evacuation and Rescue: Establish evacuation plans and procedures to access professional medical assistance or evacuate injured individuals from remote or wilderness locations, using signaling devices, communication tools and emergency contacts as appropriate to coordinate rescue efforts effectively and request professional medical care if needed.

Acquiring wilderness first aid skills and knowledge enables individuals to effectively respond to medical emergencies in remote or wilderness environments and offer lifesaving care, increasing safety and survival in such settings. Foraging for Edible Plants and Berries:

Foraging edible plants and berries in the wilderness is an invaluable skill to have for survival in an emergency. Learn to identify those locals to your region such as wild greens, fruits, nuts, and roots - research plant identification guides or take classes and workshops on local flora to expand your knowledge. Exercise caution by not eating unknown species - potentially toxic species could accidentally be consumed unwittingly! Engage in sustainable foraging practices by harvesting responsibly while leaving enough for wildlife use and future development of the plants you harvest.

## Hunting and Trapping Techniques and Tools

Hunting and trapping offer survivalists' opportunities to procure protein-rich food sources through hunting. Learn techniques like stalking, tracking, baiting, and setting traps efficiently; get acquainted with local wildlife species' behaviors and habitats so you have increased odds of success when setting traps; acquire essential hunting/trapping tools like firearms bows arrows snares traps etc.; ensure compliance with any hunting regulations/ethical guidelines to support ethical hunting practices that encourage responsible and sustainable harvests.

## Fishing: Tactics for Varying Environments

Fishing can provide reliable food sources in wilderness environments with access to bodies of water such as rivers, lakes, and streams. Learn fishing techniques such as angling, netting, and trapping to successfully catch fish; construct gear such as rods, lines hooks nets traps using natural materials or makeshift tools; adapt fishing techniques/gear to different environments/conditions such as freshwater/saltwater fishing/ shallow/deep waters fishing methods/ stationary vs mobile fishing methodologies etc.

## Building and Utilizing Primitive Tools: Spears and Arrows

Building and using primitive weapons such as spears and arrows are indispensable in survival situations for hunting, fishing, and self-defense. By learning basic woodworking and tool making techniques you can craft primitive weapons out of natural materials like wood, bone, stone, and animal parts - perfect for crafting primitive weaponry! Try your skills out hunting fishing crafting tasks to develop more skills efficiently while experimenting with designs materials techniques to produce tools designed specifically to aid wilderness survival.

## Weather Prediction and Animal Behavior

Weather prediction skills can help outdoor adventurers anticipate and adapt to rapidly evolving climate conditions more accurately in the wilderness. You should learn to read weather patterns, cloud formations, wind directions and barometric pressure changes accurately to predict upcoming weather events such as thunderstorms. Animal behavior such as birds flying low or seeking shelter could indicate impending storms or changes to atmospheric conditions - also keep an eye out for natural indicators like sky color change, smell in the air as well as insect and animal behaviors to accurately gauge these shifts in weather patterns and predict changes accurately!

Mastery of skills such as foraging for edible plants and berries, hunting, and trapping for meat, fishing for it, and signaling rescue are essential skills needed for wilderness survival. By honing these abilities individuals can increase their self-reliance, resilience and adaptability when living environments outdoors.

## DIY Project: Wilderness Survival Kit: Essential Gear for Outdoor Adventures

**Materials:**

- Water bottle or canteen
- Multi-tool or Swiss Army knife
- Fire starter (lighter, waterproof matches, or flint and steel)
- Emergency shelter (tent, tarp, or emergency blanket)

- Compass and map of the area
- First aid kit with basic medical supplies
- High-energy snacks (nuts, energy bars, dried fruits)
- Whistle or signaling device.
- Cordage (paracord or nylon rope)
- Waterproof container for storing important documents and supplies.
- Headlamp or flashlight with extra batteries
- Sun protection (hat, sunglasses, sunscreen)
- Navigation tools (GPS device or smartphone with GPS app)
- Water purification tablets or filter
- Pocket-sized survival guide or manual
- Thermal blanket or sleeping bag liner.
- Insect repellent
- Personal hygiene items (toilet paper, hand sanitizer)
- Fishing kit (line, hooks, and sinkers)
- Snare wire or small animal traps.

**Procedure:**

- Start by gathering all the materials listed above into a compact and lightweight container such as a backpack or waterproof bag, organizing each item so it's easily accessible and comfortable to carry around.
- Clean drinking water is essential to survival in any wilderness setting; fill up your bottle or canteen before heading into the wilderness using purification tablets or filters if available to ensure its safe consumption.
- Multi-tools (aka Swiss Army knives) can be an incredibly handy and multifaceted tool, used for tasks as diverse as cutting, slicing, and opening cans. Be sure that yours remains sharp and in top condition for maximum use!
- Fire is essential to human survival - from warmth, cooking and signaling for help, it plays an integral part. Be sure to pack an effective fire starter such as a lighter, waterproof matches or even flint and steel; add cotton balls soaked with petroleum jelly as tinder material for easier ignition.
- Bring along an emergency shelter like a tent, tarp, or blanket for protection from weather elements like rain, wind, or extreme temperatures. Set it up quickly if required as protection.
- Always keep a compass and map handy to help navigate your surroundings, becoming familiar with terrain features and landmarks to prevent getting disoriented or lost.
- An emergency first aid kit should include bandages, gauze pads, antiseptic wipes, pain relievers, tweezers, and any necessary prescription medicines - essential items when out in nature!
- Pack high-energy snacks to stay fueled on outdoor adventures, like nuts, energy bars and dried fruits - they're lightweight yet provide quick energy sources!

- An emergency signal device such as a whistle is an invaluable way to attract attention and signal for help should you become lost or distressed. Three short blasts are accepted as the universal distress sign.
- Cordage such as paracord or nylon rope can be invaluable tools in building shelters, securing gear, and accomplishing other tasks. Carry multiple lengths with various thicknesses for added versatility.
- Keep important documents like identity cards, emergency contacts lists and insurance info safe from water damage by keeping them in a waterproof container.
- An extra battery-equipped headlamp or flashlight is vital when traveling in low light environments or dealing with emergencies at nighttime.
- Put yourself and others under protection from harmful UV rays by wearing a hat, sunglasses, and sunscreen regularly - especially after sweating or swimming. Reapply sunscreen as necessary, particularly after exertion such as exercising.
- Navigation devices such as GPS devices or smartphones with built-in navigation apps are great ways to stay oriented when traveling abroad, should you become lost. Just make sure that you bring extra batteries or portable chargers for any electronic devices you bring along!
- If your water supply needs replenishing, use your fishing kit to catch fish from nearby waters and set traps or snares for smaller creatures if necessary - these methods could provide food as well.

**As part of good hygiene practices to avoid illness and infection, always bring toilet paper, hand sanitizer and waste management supplies for cleanliness, disposing of your waste correctly to minimize environmental impacts.**

# BOOK 16: COMMUNICATION TACTICS FOR CRISIS

Establishing communication protocols and channels are paramount for effective emergency communications. Create an elaborate communication plan outlining protocols for initiating, transmitting, receiving, identifying primary and secondary channels like radios phones satellite communication platforms, internet-based platforms as primary or secondary communication modes respectively and then establish clear procedures for activating these channels, disseminating information coordinating response efforts as well as training community members / stakeholders regarding these protocols in advance of emergencies to ensure reliable responses during these situations.

## Emergency Equipment: Radios and Phones

Emergency equipment, like radios and phones, plays an essential part in keeping people communicating during crises. For maximum effectiveness during emergency scenarios, invest in reliable two-way radios with long range capabilities and multiple channels allowing communication among individuals, teams and command centers - keep these properly maintained, charged, programmed with the relevant frequencies/channels/frequency bands etc. - along with mobile phones equipped with adequate battery life, network coverage and emergency contact lists; make calls using them or text them immediately in times of distress!

## Code Words and Signals to Establish Secure Communication

Code words and signals provide secure communication in situations requiring confidentiality and security. Use predetermined code words, phrases, and signals to convey sensitive data securely while also coordinating actions and authenticating messages securely. Train personnel on how best to interpret these code words/signals to minimize interception or compromise from unintended parties.

## Setting Up Neighborhood Networks to Facilitate Communication

Building neighborhood networks to enhance community resilience and coordination during emergencies is an effective way to strengthen resilience and coordination within localities. Establish communication networks within neighborhoods, communities, or localities so residents can share information, allocate resources efficiently and mutually assist one another. Designate communication hubs as leaders or volunteers responsible for disseminating alerts or instructions among community members; promote participation from neighbors by inviting participation or collaboration on joint efforts to strengthen social ties, build trust among them all while strengthening communication resilience within your locality.

## Utilizing Social Media for Crisis Communication

Social media platforms provide invaluable crisis communication platforms, enabling individuals and organizations to rapidly distribute real-time updates and alerts. Establish official accounts/channels on these social platforms for emergency management agencies, local authorities, community organizations etc. so they can share crucial info when crises strike; monitor these channels regularly for relevant updates/feedback/enquiries from community members - responding promptly with timely assistance to address help as soon as they arise.

## Emergency Communications with Mobile Alerts, Updates and Apps

Mobile alerts, updates and apps provide essential tools for providing timely and relevant information in emergencies. Subscribe to alert systems provided by government agencies, local authorities, and mobile carriers to stay abreast of emergencies, severe weather, and other critical events; download emergency communication apps designed for response that provide access to essential resources, resources, and communication channels in one easy platform; subscribe to emergency notification services that alert on weather emergencies provided by government bodies & carriers as well - which provide alerts.

Establishing communication protocols and channels, taking advantage of emergency equipment, using code words and signals for emergency purposes, setting up neighborhood networks, exploiting social media and employing mobile alerts/apps are essential strategies for aiding effective communication during an emergency. Communities that adopt these measures will find improved resilience, coordination, and response capabilities during times of crises.

## Contact Lists of Family, Friends, and Essential Services

Maintaining updated contact lists is critical to effective emergency communication and coordination. Store contact info in one central place (like a physical notebook or digital document), including details like these for effective emergency planning:

**Family Members:** Make an inventory of immediate family members such as parents, siblings, spouses, and children; this should include medical information or special requirements of each member of your immediate family unit.

Friends and Relatives: Be sure to list contact info for close friends, relatives and neighbors that could offer emergency support during times of trouble or require special skills or resources from you to assist. Specify their relationship to you as well as any specific qualities they possess that could assist them during these trying times.

Emergency Services: Keep contact numbers for local police agencies and municipal services handy to dial for emergencies like police, fire, ambulance, and poison control centers. Furthermore, include non-emergency contact numbers that you might use if there's ever any need.

Medical Providers: Record all contact details for healthcare providers such as physicians, hospitals, and medical facilities that you or a member of your family use for care; including emergency contact numbers and after-hours care instructions.

Utilities and Services: Provide contact numbers of utility providers such as electric, gas and water providers as well as internet/phone service providers as well as plumbing/HVAC maintenance companies for reporting service outages or emergencies.

Maintain a regularly updated contact list to ensure accuracy and ease of accessibility in an emergency, sharing copies with family and trusted contacts for quick reference.

## Licensing and Operating Ham Radio

Ham radio licensing and operation provide reliable communication for individuals during emergencies, natural disasters, and other times when traditional communication methods may become inaccessible or unreliable. To obtain and legally operate HAM equipment, follow these steps.

Study for the License Exam: Develop your understanding of HAM radio regulations, operating procedures, and technical concepts by studying for your license exam. There are study guides, online courses, and practice exams available to you as tools to assist your study efforts and prepare you for this important test.

Take Your License Exam: Schedule and take your HAM radio license exam administered by volunteer examiners authorized by the Federal Communications Commission. It consists of multiple-choice questions covering topics like regulations, operating practices, and technical knowledge.

Your License: After passing the exam, submit your paperwork and application to either the FCC or an organization such as ARRL to acquire your HAM radio license. Your operating privileges and frequency allocations depend upon which class your license falls into (e.g., Technician, General or Extra).

Acquire Equipment: Purchase the essential HAM radio equipment such as transceivers (radio), antennae, power sources and all required accessories for operation of handheld or base stations depending on your license class and intended operating activities, such as handheld radios for portable use or base stations for home usage.

Attain Operating Procedures: Take time to familiarize yourself with HAM radio's operating procedures, etiquette, and best practices for effective communication. Join local clubs, nets and activities that offer these training opportunities so that you may gain experience under experienced operators' tutelage.

By becoming licensed HAM radio operators and operating the equipment responsibly, individuals can contribute to emergency communication networks, stay informed during crises, and offer help and aid to others who require it.

## Communication With Local Authorities: Establishing Contact

Communicating effectively with local authorities is integral for receiving timely updates and assistance during emergencies or crisis situations. Follow these steps to initiate communication between them:

- Establish Key Contacts: Conduct extensive research to locate key personnel at local government agencies, emergency management offices, law enforcement departments and municipal services responsible for emergency response and communication.
- Make an Emergency Contact List: Create and save phone numbers, emails addresses and any additional contact info needed for local authorities in an emergency contact list. Incorporate non-emergency numbers for general inquiries or administrative purposes as needed.
- Follow Official Channels: Keep tabs on official communication platforms used by local governments for providing emergency updates, such as government websites, social media accounts, emergency alert systems and news outlets.

- Register for Alerts: Signing up with local emergency alert systems and notification services provides timely alerts regarding potential hazards, severe weather events, evacuation orders and other vital details that will protect life and property in emergencies.
- Take Part in Community Programs: Get involved with community programs, workshops and outreach initiatives organized by local authorities that aim to foster emergency preparedness, communication, and cooperation between residents.
- Establish Communication Protocols: Make plans and set procedures in case of emergencies, reports incidents, seeking help and seeking information. Make these protocols known among family and household members so they may refer to them when needed.
- Individuals can increase their preparedness and resilience in times of emergency by communicating with local authorities and keeping informed on emergency procedures and resources and staying in communication. By communicating and staying up to date about them, people will increase their preparedness for unexpected crises or disasters.
- Text Messaging Systems as Emergency Communication Text messaging systems offer an efficient and dependable means of emergency communication that allows individuals to quickly send and receive critical updates via mobile phones and devices via text messages. When applied for use during emergencies:
- **Register for Alert Services: Register with text-based emergency alert systems and notification services provided by local governments, government agencies and emergency management organizations** in your area. Register your mobile phone number with them so you'll start receiving alerts in your neighborhood!
- Stay Informed: Stay up to date on local emergency procedures, hazards, and evacuation plans by monitoring text alerts from official sources regularly. Be mindful of instructions, warnings and safety recommendations provided in emergency messages.
- Spread Awareness: Encourage family, friends, and neighbors to register for text-based emergency alert systems as a means of staying informed and prepared during emergencies. Distribute information regarding how they can sign up for alert services as well as accessing emergency resources and assistance services.
- Prepare Emergency Messages and Communication Templates in Advance: Make plans now for alerting, updating, and instructing during emergencies by creating emergency messaging templates in advance for alerting, updating, and instructing individuals of what's happening, including essential details on its nature as well as safety recommendations, evacuation orders and contact info of local authorities and emergency services.
- Verify Communication Systems: Conduct regular tests on text messaging systems and emergency communication channels to make sure they can reliably send out emergency alerts in times of crises, with messages reaching intended recipients promptly and accurately.

By employing text messaging systems for emergency communication purposes, individuals can quickly receive alerts about crises or disasters, share essential data quickly and coordinate response efforts efficiently in an emergency.

- Emergency Broadcast Messages: Preparing and Broadcasting Preparing and broadcasting emergency broadcast messages effectively are of critical importance during times of emergencies, natural disasters, or crisis situations. To do this effectively:
- Construct Message Contents: Craft emergency alert messages containing essential details regarding the nature of emergency, safety instructions, evacuation orders and shelter locations as well as contact info for local authorities and emergency services.
- Adhere to Protocols: Adherence to emergency messaging protocols set out by government agencies, emergency management offices and communication authorities must take priority when broadcasting emergency alert messages and broadcasts in emergencies. Make sure messages comply with legal requirements, accessibility standards and cultural sensitivities guidelines as established.
- Apply Multiple Channels and Platforms: To reach as broad an audience as possible with emergency messages, use various communication platforms including radio, television, social media platforms like Facebook or Instagram; texting services; email services; siren systems; public address systems to effectively deliver emergency updates to residents, visitors, and stakeholders alike.
- Coordinate with Authorities: Collaborate with local authorities, emergency responders and government agencies responsible for emergency management to ensure accurate, timely, and coordinated messaging efforts. Partner with relevant stakeholders to develop customized messaging strategies and response plans tailored specifically for individual emergencies or scenarios.
- Test and Update Systems: Regularly conduct tests of emergency broadcast systems and communication channels to make sure they can deliver messages effectively during emergencies. Engage in drills, exercises, and simulations to practice emergency messaging procedures as well as identify areas that could be used for improvement.

## DIY Project: Emergency Communication Kit: Staying Connected During Crisis

**Materials:**

- Portable two-way radios (walkie-talkies)
- Hand-crank or solar-powered emergency radio
- Signal mirror.
- Whistle
- Cell phone with spare battery or portable charger
- Prepaid phone card
- Paper and pens
- List of emergency contacts
- Map of the area
- Personal locator beacon (PLB) or satellite messenger device
- Bullhorn or megaphone
- Glow sticks or chem lights.
- Flares or smoke signals
- Emergency communication plan (printed copy)

**Procedure:**

- Begin by gathering all the materials listed above into an easily transportable container like a backpack or waterproof bag and making sure every item is working efficiently in case of an emergency. Make sure your supplies can easily be found when needed!
- Portable two-way radios, more commonly referred to as walkie-talkies, provide essential short range communication tools between individuals or groups. Make sure every member of your party owns one and is set on the same channel to facilitate quick dialogue.
- An emergency radio powered by either a hand-crank or solar energy can come in very handy when an unexpected disaster arises, providing weather updates, emergency broadcasts, and important information in times of need. Make sure it is tuned into local emergency frequencies for optimal operation - test it frequently to make sure everything works smoothly!
- Signal mirrors can help attract rescuers or aircraft. Take some practice using one to communicate a need for help and familiarize yourself with its use and proper technique.
- Always carry one with you as it provides an effective means of signaling for help if you become lost or injured, using it if appropriate to travel long distances.
- Staying connected during an unexpected power outage requires keeping your cell phone fully charged, with spare batteries or portable chargers at hand just in case the power runs out. As an extra safeguard, consider purchasing a prepaid phone card as backup should cell service become unavailable.
- Pack paper and pens so you can leave notes or messages for others, particularly if your group requires directions or information from you.
- Create an emergency contact list encompassing family, friends, and local authorities; keep it handy and give copies to each member of your group.
- Keep a map handy of your surroundings to get familiarized with both terrain and nearby landmarks - then use this map as a navigation aid and to notify rescuers if required.
- Consider investing in a personal locator beacon (PLB) or satellite messenger device when undertaking remote travel or outdoor excursions, as these devices can send distress signals and GPS coordinates directly to emergency responders should an incident arise.
- Bullhorns or megaphones are great tools to amplify your voice during times of crises or emergencies, providing communication among a larger group.
- Glow sticks/chem lights can help mark your location or signal for help during low light situations, so keep a supply on hand and activate, as necessary.
- Flares or smoke signals can be an effective way to attract rescuers or passing aircraft during both daylight and nighttime hours, providing essential emergency alerting capabilities. When employing these devices, it's imperative that proper safety protocols are strictly observed so only use these in critical emergency scenarios.
- Create and distribute copies of an emergency communication plan among members of your group, reviewing it periodically and making updates as necessary so everyone knows what steps to take if an incident arises.

# BOOK 17: THE PREPPER'S AID GUIDE

Examining a situation starts by assuring both responders and victims are safe at the scene of an emergency, including potential hazards like fire, traffic congestion or hazardous substances. After making sure it is secure for everyone on-scene, conduct triage to prioritize care according to severity. Triage involves quickly assessing victims for life-threatening injuries which require immediate care immediately while those suffering serious injuries but who have more minor wounds may aid or wait until care can arrive for treatment or provide aid until then.

## Basic Anatomy and Physiology for First Aid

Understanding basic anatomy and physiology is integral for providing effective first aid. Familiarize yourself with major body systems, organs, and vital signs. Learn to recognize common medical emergencies like heart attack, stroke, allergic reactions, or respiratory distress symptoms and understand their responses more quickly when providing first aid treatment to victims of injuries or illnesses. Understanding body responses allows you to treat victims more confidently.

## CPR (Cardiopulmonary Resuscitation) and AED Use

CPR and AED use are crucial skills for quickly responding to cardiac arrest emergencies. Learn to perform high-quality CPR on adults, children, and infant victims alike while simultaneously using an automated external defibrillator (AED). AED training courses teach techniques for using AED shocks on victims with sudden cardiac arrest as well as techniques for activating emergency medical services and coordinating care among responders.

## Treating Severe Bleeding and Wound Management

Treating severe bleeding and managing wounds requires swift, decisive actions to stop further injury or infection. Learn to apply direct pressure, elevate injured limbs, and use tourniquets or pressure dressings for tourniquet-controlled pressure dressings for rapid wound closure and pressure dressing used to control bleeding. Clean wounds using sterile saline solution (or clean water) to flush away debris that increases infection risks; apply appropriate dressings such as bandages or sutures that will protect and promote healing as soon as possible while monitoring for signs of shock as necessary and providing supportive care accordingly.

## Splinting Fractures and Immobilizing Injuries

Splinting fractures and immobilizing injuries help protect bones, muscles, and soft tissues against further injury. You will learn how to assess for fractures, dislocations, sprains and stabilize injured limbs with either homemade or commercial splints; apply padding around injured area for support and cushioning while secure the splint using bandages, wraps or tape so circulation doesn't become compromised; immobilize injured limbs in positions which provide comfort, so pain reduction occurs more easily while transport takes place without further traumas.

Mastery of essential skills such as assessing situations, comprehending basic anatomy and physiology, performing CPR with an AED device, treating severe bleeding/wound management as well as splinting fractures/immobilizing injuries is crucial in providing first aid during emergency situations. By becoming skilled first aiders themselves and becoming confident responders capable of making a difference by saving lives or lessening injuries caused by illnesses/injuries/or traumas/illnesses incurred.

## Acknowledging and Addressing Shock

Shock occurs when our vital organs do not receive enough oxygen and nutrients due to inadequate blood flow, leading to rapid pulse, shallow breathing, clammy skin, confusion, and unconsciousness. To effectively treat shock in an emergency situation, first ensure scene safety, then evaluate airway, breathing and circulation (ABCs) needs of patient in question before lying patient down and elevating legs to increase circulation to vital organs; blanket up patient warm without overheating; monitor vital signs closely as soon as help arrives, while providing emotional and reassurance while waiting on medical assistance as quickly as possible while providing emotional and supportive assistance while waiting on medical professionals to arrive on scene.

## Approaches for Treating Burns at Different Degrees

Burns can result from exposure to heat, chemicals, electricity, or radiation and are classified into three degrees by their severity: first-degree burns are limited to the outer layers of skin and exhibit redness, pain, and mild swelling while second-degree burns extend deeper within, leading to blistering severe pain swelling or blackening charring and may appear white blackened or even charred. To treat burns effectively use running water on affected area for cooling purpose & cover burn with clean dry dressing before seeking medical assistance for severe burns such as those involving large areas of body such as Face hands feet genitalia.

## Coughing

Choking occurs when airway obstruction prevents breathing. Common indicators include inability to speak, cough, or breathe normally; clutching of throat; and a bluish tint on lips or skin. To perform the Heimlich maneuver on an unconscious victim who is conscious, stand behind them, wrap your arms around their waists, make a fist using one hand then place slightly above navel but below ribcage before grasping fist sharply then thrusting inward and upward with another. Repeat these thrusts until either object dislodged from their airway or unconscious victim becomes unconscious; otherwise perform back blows and chest thrusts instead.

## Treating Head Injuries and Concussions

Head injuries such as concussions can result from falls, blows to the head or motor vehicle accidents. Common symptoms of head injury may include headache, dizziness, nausea, confusion, blurred vision, or loss of consciousness. If you suspect head trauma, assess the level of consciousness pupil size reaction motor function. Provide rest to keep person still while monitoring vital signs mental status Vitals status should always be monitored or seek medical assistance immediately for severe headache or loss of consciousness requiring hospitalization or loss of consciousness for which medical assistance must be sought immediately (vomiting consciousness), loss of consciousness symptoms persist for any length of time period or exhibit signs worsening neurological function deterioration due to injury if symptoms exist such as vomiting seizures severe headache or worsening neurological function or symptoms appear that worsen neurological function worsens significantly).

## Allergic Reactions and Anaphylaxis

Allergic reactions occur when our immune systems overreact to foreign substances like food, medication, or insect venom. Anaphylaxis is a potentially life-threatening allergic response which needs immediate medical intervention; symptoms may include hives, itching, facial and throat swelling as well as difficulty breathing, rapid heartbeat, and loss of consciousness. When administering Epinephrine with an autoinjector such as

EpiPen, call 911 immediately, administer Epinephrine immediately if available and monitor airway breathing circulation and airway breathing before providing CPR if required, if necessary, reassurance while keeping legs elevated to improve blood flow and improve circulation.

## DIY Project: Prepper's Comprehensive First Aid Kit: Essential Supplies for Emergency Aid

**Materials:**

- Adhesive bandages (various sizes)
- Sterile gauze pads
- Medical adhesive tape
- Antiseptic wipes or solution
- Alcohol pads
- Hydrogen peroxide
- Triple antibiotic ointment
- Tweezers
- Scissors
- Instant cold packs
- Elastic bandages (ACE bandages)
- Sterile saline solution
- Oral thermometer
- CPR mask or face shield
- Disposable gloves
- Tourniquet
- Splinting materials (popsicle sticks, SAM splints)
- Irrigation syringe
- Burn gel or burn dressings.
- Emergency blanket

**Procedure:**

- Begin by collecting all the materials listed above and organizing them into an easily accessible, durable container or bag. In case of emergency situations, ensure it can accommodate immediate access.
- Adhesive bandages are indispensable tools for covering minor cuts, scrapes, and blisters. Choose different sizes to ensure they fit various wounds or body parts.
- Sterile gauze pads can help disinfect and dress larger wounds safely and sterility. Select pads individually packaged to maintain their sterility.
- Medical adhesive tape is essential to secure dressings and bandages in place, particularly for individuals with sensitive skin. Be sure to include hypoallergenic rolls of tape for those prone to allergies.
- Antiseptic wipes or solutions can be used to clean wounds and avoid infections, while alcohol pads provide extra disinfection prior to administering injections or medical procedures.

- Hydrogen peroxide is a mild antiseptic used for cleaning wounds and clearing debris away, however when applied directly onto open wounds it could potentially sting or cause irritation. Use caution if applying directly onto open wounds as it could cause discomfort and cause further harm.
- Triple antibiotic ointment helps protect and heal minor cuts and abrasions quickly and painlessly. Apply a thin layer to wounds before covering them with bandages for the best results.
- Tweezers can help remove splinters, thorns, and other foreign objects from the skin with precision extraction. Choose a pair with fine tips for precision use.
- Scissors are essential tools in an emergency, especially for cutting bandages, tape, and clothing off wounds safely and quickly. Be sure to pack several pairs with blunt tips so they do not cause unnecessary injuries when cutting close-cut clothing away.
- Instant cold packs offer immediate comfort from sprains, strains, and minor burns. Simply activate one by compressing or shaking to activate its cooling action and experience instantaneous relief!
- Elastic bandages (ACE bandages) can provide support and compression when applied over injured joints to provide support and comfort.
- Sterile saline solution can be used to quickly rinse wounds, eyes and mucous membranes of debris and infection. Pack an individual dose via small bottle or ampoules for single use applications.
- An oral thermometer is essential to monitoring body temperature and identifying fever or **hypothermia, so selecting one with digital technology for accurate yet easy-to-read result**s.
- CPR masks or face shields provide extra protection while performing cardiopulmonary resuscitation (CPR) in case someone stops breathing or doesn't have heartbeats. Make sure your first aid kit includes one and familiarize yourself with CPR procedures before an emergency arises.
- Disposable gloves are vital tools in protecting yourself against infection during medical procedures and protecting the hands from bodily fluids, so pack several pairs in different sizes to accommodate different users.
- Tourniquets can help control severe bleeding from extremities in emergency situations. Learn to properly apply one, using it only after direct pressure has failed or other techniques have proven ineffective.
- Popsicle sticks or SAM splints can be useful tools in immobilizing fractures and stabilizing injured limbs, providing immobility during healing processes and relieving pain associated with injuries to bones or muscles. Pack several sizes of these splints to accommodate for all injuries sustained during an incident.
- An irrigation syringe is useful for flushing away debris from wounds and irrigating eyes or mucous membranes, using either sterile saline solution or fresh water as the means. Use it regularly until all affected areas have been thoroughly cleansed of dirt.
- Burn gel or dressings provide temporary relief and aid in the prevention of infection for minor burn injuries. Be sure to follow all the manufacturer's instructions regarding use.
- Emergency blankets (commonly referred to as space or thermal blankets) help maintain body heat during inclement weather and help avoid hypothermia - providing invaluable warmth in harsh climate conditions. Carry one in your first aid kit as an additional safety measure should an emergency or survival situation arise.

# BOOK 18: CITY SURVIVAL TACTICS

Understanding urban risks and threats is vital for effective urban survival preparation. Urban environments present unique challenges like crime, violence, traffic congestion, pollution, infrastructure failures and natural disasters - so conducting a risk analysis to identify hazards in your local area (like crime hotspots, industrial sites flood zones earthquake-prone regions etc.). Be informed on local emergency plans evacuation routes shelter locations to develop strategies which mitigate risks efficiently while responding effectively.

## Building Urban Navigation Skills with Maps, GPS, and Landmarks

Urban navigation skills are critical when dealing with emergencies or crises in city environments. Familiarize yourself with maps, street layouts and landmarks of your locality to aid navigation and orientation, then utilize GPS devices, smartphone apps or navigation tools in real time for route planning, finding alternative paths or tracking your location - use them even while on foot! Learn street signs interpretation along with landmark interpretation as you become comfortable navigating urban environments efficiently and safely - take walks, bike rides or drive excursions regularly to build confidence and become comfortable navigating urban environments! Practicing urban navigation through regular walks, bike rides or driving excursions can build familiarity and gain comfortability with surroundings while increasing confidence by becoming familiarized with them over time and practiced mastered over time!

## Building Emergency Kits for Urban Survival

Establishing emergency kits for urban survival is vitally important to being ready in case of unexpected disasters and emergencies. Your emergency kits should include essential items like water, food, first aid supplies, flashlights, batteries, multi-tool, whistle, portable radio, cash emergency funds and important documents - these should all be customized specifically according to the circumstances and duration of emergency events in which they'll be used in. Keep them easily accessible at home, work or vehicle and periodically review and update them so they're fully functional at any given moment in time.

## Urban Environments for Securing Shelter

Securing shelter in urban environments often involves seeking refuge in buildings, parks, public facilities or emergency shelters during crises or disasters. Locate nearby community centers, schools, churches, or government buildings which could serve as potential shelter options; also familiarize yourself with evacuation routes, assembly points and designated shelter locations in your neighborhood as well as emergency protocols and access procedures at public shelters as well as potential alternative locations such as friends' or relatives' homes, hotels, or commercial properties in case such an event should arise.

## Locating and Purifying Water in Urban Settings

Finding and purifying water sources in urban settings are an integral component of providing access to safe drinking water in times of emergencies or interruptions to water supplies. Look out for local taps, faucets, fountains or hydrants offering water sources in urban settings; carry portable filters or purification tablets or UV sterilizers (to treat dubious sources); exercise care when collecting urban sources so as to avoid contaminants like pollutants, chemicals or sewage contamination that might compromise its purity and potability; exercise caution when collecting such water sources in order to safeguard ensuring safe consumption during emergencies or disruptions to ensure safe access during emergencies or disruptions to supply chains.

## Gathering Food and Supplies in Urban Areas

Urban food gathering requires scavenging, foraging, bartering, or accessing local resources during emergencies or disruptions in supply chains. Locate nearby stores or warehouses which might contain emergency provisions of food, water or essential supplies needed in an emergency, stockpile non-perishable foodstuffs like canned goods snacks rations in your home or shelter location for immediate provisioning needs, explore opportunities for urban foraging for edible plants fruits wild foods in parks gardens green spaces etc. and establish connections with neighbors or mutual aid groups so as to share resources skills and support each other during times of crises.

## Utilizing Public Transportation and Alternative Routes

At times of urban crisis, public transportation can provide invaluable aid for evacuation or mobility. Familiarize yourself with bus, subway and train routes as well as their schedules and emergency protocols before an urban crisis hits; identify alternate modes of transport should a disruption arise or overcrowding occurs; walking/bicycling as backup solutions or sharing mobility services may be more suitable alternatives; stay updated through official channels as well as local news sources about any service updates/changes and keep yourself abreast of service updates/changes as backup options!

## Navigating Urban Traffic and Roadblocks

Navigating urban traffic and roadblocks successfully takes adaptability and resourcefulness. Utilize traffic reports, real-time navigation apps, or alternative routes and side streets to avoid congestion or road closures; identify side streets to bypass any jams; remain patient while keeping a safe distance between vehicles to reduce accident risks; be flexible enough to adjust plans based on changing conditions or unexpected incidents if necessary; be ready for sudden roadside emergencies that require your attention or adjust plans and routes as required - especially during times of high volumes or extreme traffic events!

## Securing Your Apartment or Home in an Urban Crisis

Securing your apartment or home during urban crises is vitally important for personal and property protection. Secure the doors and windows using sturdy locks, security bars or shatter-resistant film to increase protection. Keep emergency supplies such as food, water flashlights and first aid kits readily accessible as well as communication and evacuation plans with household members or neighbors easily accessible. Consider installing security cameras or alarms as added measures of defense while engaging in community watch programs to provide mutual support and vigilance.

## Maintaining Communication in Urban Areas

Maintaining communication in urban environments requires reliable and versatile tools. Carry a charged cell phone with emergency contacts programmed, portable chargers or solar chargers available as backup power sources and walkie talkies that offer short range communication for short range communication such as family or neighbors; consider investing in two-way radios/walkie talkies as short range solutions; establish meeting points or check-in protocols should communication breakdown, stay updated via official channels as well as community networks on updates related to emergencies if separate from family.

## Avoid Crowded and Dangerous Areas

Avoiding crowds and dangerous areas during an urban crisis is paramount to personal safety. Rely on your intuition if a situation feels unsafe or threatening; plan travel routes that reduce exposure to high-traffic areas that might spark violence or unrest, avoid confrontation with individuals or groups and seek law enforcement or authority assistance as necessary - staying connected with trusted contacts while remaining aware of changing circumstances is also key for survival.

Overall, public transit use, navigation of traffic flow, home security measures, waste management practices, maintaining communication lines and avoiding crowds are key strategies for urban survival preparation. By actively taking these measures into consideration and acting upon them during emergencies or crises situations, individuals can increase resilience and safety within urban environments.

## DIY Project: Urban Survival Kit: Navigating City Environments in Crisis

**Materials:**

- Backpack or messenger bag.
- Water bottles or hydration reservoir
- Non-perishable food items (energy bars, canned goods, etc.)
- Multi-tool or Swiss Army knife
- **Flashlight or headlamp with extra batteries**
- Portable phone charger or power bank
- Map of the city
- Compass
- Whistle or signaling device.
- Cash (small bills and coins)
- First aid kit with essential medical supplies
- Personal hygiene items (toilet paper, hand sanitizer, etc.)
- Lightweight rain poncho or emergency shelter
- Pocket-sized umbrella
- Protective face mask
- Gloves
- Duct tape.
- Ziplock bags or waterproof pouches for storing important documents.
- Notepad and pen
- Emergency contact list

**Procedure:**

- Start by selecting a sturdy yet comfortable backpack or messenger bag to carry your urban survival kit. Make sure it features multiple compartments, so your supplies are organized easily while giving easy access to essential items.
- Water is essential to survival; be sure to pack multiple water bottles or hydration reservoirs in case of emergency situations. Aim to carry at least one liter per person per day.

- Choose non-perishable food items like energy bars, canned goods and snacks which require no preparation or refrigeration in your emergency food supplies to sustain energy and sustain vitality during a crisis. Select foods high in protein and calories to maintain your strength during an unexpected crisis.
- Multi-tools or Swiss Army knives are versatile tools designed to accomplish various tasks such as cutting, slicing, and opening cans. Be sure yours includes essential tools like knives, screwdrivers, and scissors.
- Keep a flashlight or headlamp equipped with additional batteries at hand in your bag to illuminate low light conditions and power outages and be ready for emergencies by keeping it within easy access in an easily identifiable place.
- An external power bank or portable charger can keep your electronic devices powered up during an emergency and enable access to crucial information that could save lives.
- Acquaint yourself with the layout of the city by carrying a map and compass. Use them to navigate your surroundings and devise alternate routes in case of road closures or congestion.
- Carry a whistle or other always signaling device with you so you can draw attention and request help if lost or in distress; three short blasts are the universal signal for distress.
- Your urban survival kit should contain small bills and coins in case electronic payment systems become inaccessible during an emergency.
- Create a first aid kit stocked with necessary medical supplies like bandages, gauze pads, antiseptic wipes, pain relievers and any necessary prescription medicines.
- Include personal hygiene products, like toilet paper, hand sanitizer and feminine hygiene items as essential emergency necessities to remain comfortable during an unexpected disaster situation.
- An emergency shelter, like a rain poncho or umbrella can offer protection in inclement weather or unexpected situations.
- Take precaution by carrying a compact umbrella with you when venturing out on city streets to provide shade from rain or sun exposure.
- Protect yourself against airborne pollutants or infectious diseases by wearing a face mask when necessary.
- Be sure to include gloves in your urban survival kit to protect your hands during emergency situations or when handling hazardous materials or debris.
- Duct tape is an extremely handy solution that can quickly repair damage, secure items, or provide emergency solutions in times of trouble.
- Use Ziplock bags or waterproof pouches to safely store important documents such as identification cards, insurance papers and emergency contacts - to ensure their protection and avoid misplacing them or losing them during travel.
- Always carry a notepad and pen with you so you can quickly take notes, record pertinent information, or leave messages for others.
- Establish an emergency contact list consisting of family, friends, and local authorities for easy reference during times of trouble or uncertainty. Keep copies in your urban survival kit as well as sharing copies with trusted individuals.

Keep your urban survival kit updated to remain relevant, in good condition, and fresh by periodically reviewing and revising its contents regularly to rotate perishable items, replace expired ones as appropriate, and adjust according to changes in needs or circumstances.

# BOOK 19: SHELTER ESSENTIALS: MAINTAINING CLEANLINESS AND SAFETY

The cleanliness of shelter surfaces is vital in creating a healthy living environment. Establish a cleaning schedule to remove dust, dirt and debris from floors, walls and other surfaces with appropriate cleaners and equipment; pay special attention to high traffic areas like kitchens or bathrooms where bacteria or germs could gather; inspect for signs of wear-and-tear quickly so further damage doesn't arise; make any needed repairs swiftly as issues may worsen further deterioration occurs.

## Ventilation to Prevent Mold and Mildew

Proper ventilation is vital in protecting against the development of mold and mildew that poses health risks to indoor air quality, including decreasing oxygen levels and condensation formation. To maximize airflow in bathrooms and kitchens by opening windows or using exhaust fans to remove moisture. In damp or humid areas use dehumidifiers to lower humidity levels and avoid condensation forming. Finally, regularly inspect ventilation systems, ductwork, filters for obstructions or malfunctions then clean or replace as necessary to maximize indoor air circulation and circulation.

Pest Control Measures and Entry Point Sealing: To be implemented as soon as possible to stop insects entering, seal all entry points, or take preventive steps such as sealing holes to block entryway points for them to gain entrance.

Implementing pest control measures and sealing entry points are integral for protecting shelter occupants against rodents, insects, and other vermin. Conduct regular inspections of both the exterior and interior to identify potential entry points such as cracks in walls or doors; gaps or openings between panels on floors; cracks between window panes etc. that might allow in vermin; then seal any entryways using caulk, weatherstripping or mesh screens so no pest can gain entry to invade; use traps baits or natural deterrents when necessary to control existing populations or prevent reinfestations of shelter occupants from reinfestation by controlling existing populations or reinfestation from occurring again.

## Precautions to Avoid Fire Emergencies

Fire safety precautions are vitally important in safeguarding shelter occupants from fire-related emergencies. Install smoke and carbon monoxide detectors throughout your shelter - such as bedrooms, hallways, and common spaces. Make sure they're regularly tested to make sure batteries don't run low; test regularly after installing new detector batteries to make sure everything's in working order; ensure portable fire extinguishers are easily accessible in designated spots while training occupants on how to use them effectively; devise and practice an escape route plan including multiple evacuation routes as well as designated meeting points should an incident arises involving potential meetings points - just in case something could go wrong and need help be put together quickly - in case it happens,

## Identification of Structural Weaknesses and Damage

Regular inspections are vital in detecting structural weaknesses that could compromise the safety and integrity of shelters, including cracks in walls, ceilings, floors, foundations, and structural components like beams columns and supports that could compromise its strength or collapse. Conduct thorough reviews on both interior and exterior aspects, such as walls ceilings floors foundations beams columns supports etc. to identify areas needing work such as cracks sagging deterioration cracking water damage etc. which need urgent addressing as early as possible to preventing further damages or collapse from furthering damage occurring or collapsing altogether. When needed, seek professional evaluation as repair solutions from experts for help as needed to address structural issues caused by defects.

## Installing Carbon Monoxide Detectors in Living Spaces

Carbon monoxide detectors are essential tools in the battle against this deadly gas, produced from malfunctioning or poorly ventilated fuel-burning appliances such as furnaces, stoves, and water heaters. Install carbon monoxide detectors near sleeping areas in living spaces; test regularly to make sure they're functioning as intended and follow manufacturer recommendations on placement, maintenance, and replacement to provide continuous protection from carbon monoxide poisoning.

## Reducing Accidental Incidents When Storing Hazardous Items

Secure hazardous items to reduce accidents and protect occupants of shelter facilities from harm. Care must be taken in handling and storing cleaning products, chemicals, flammable liquids, sharp objects and sharp objects in designated containers or cabinets with childproof locks or safety latches installed to block access by children or unauthorized individuals. Manufacturer guidelines on handling and storing such materials as well as proper labeling to alert potential risks should always be observed and followed when handling and storing these hazardous materials.

## Clear Away Clutter for Safe Paths

Create safe paths by clearing clutter to lower the risk of trips, falls and injuries in shelter environments. Conduct regular declutter sessions in living spaces, hallways, and walkways to remove obstacles such as furniture boxes or debris that obstruct movements as well as loose rugs cords or any other trip hazards causing trips in dimly lit environments and ensure optimal visibility while providing enough lighting in dimly lit places for improved visibility and accident avoidance. Encourage shelter occupants to maintain tidy living environments for improved accessibility as a measure towards maintaining tidy living quarters which contribute towards safety and accessibility!

## Proper Storage of Hazardous Chemicals and Materials

Proper storage of hazardous chemicals and materials is critical to avoid spills, leaks, and environmental contamination in shelter environments. Store hazardous substances in well-ventilated spaces away from heat sources, direct sunlight and materials incompatible with them; use sealed bottles, drums or bins with secure seals as storage units to avoid accidental exposure and leakage of leakage or accidental release; implement secondary containment measures (spillage trays/berms etc.) so as to contain spills without environmental contamination; train shelter occupants on safe handling practices as well as emergency response procedures when dealing with hazardous material hazards; train shelter occupants on safe handling practices/ emergency response protocols when dealing with hazardous material hazards.

## Regular Testing of Systems, Alarms and Generators

Regular testing of systems, alarms, and generators in shelter environments is vital to maintain readiness and reliability. Be sure to schedule routine inspections and maintenance checks of fire detection and suppression systems, carbon monoxide detectors, smoke alarms and emergency lighting as well as backup generators/battery backups/UPSs/uninterruptible power supply (UPS) systems to ensure their functionality in times of power outage or emergencies; change batteries as necessary while keeping records of maintenance activities for compliance and documentation purposes.

## Establish Emergency Communication Protocols

Establishing emergency communication protocols is critical to ensure timely and effective responses in crises or emergencies. Establish clear procedures for alerting shelter occupants, staff, and relevant authorities of emergencies like fires, natural disasters, or security incidents; designate primary and alternate communication channels like phone trees, text alerts or intercom systems for this purpose. Train both staff members as well as residents on how to report emergencies promptly as well as who to contact and what information will need to be shared when reporting an event or crisis.

## Implement Security Measures to Prevent Break-Ins

Implementing security measures is vital for deterring break-ins and protecting shelter occupants and property from break-ins, as it helps mitigate security risks such as inadequate lighting, unsecure entry points or no surveillance coverage. Assess security risks such as inadequate lighting levels or entry points without sufficient surveillance coverage and install and maintain systems such as locks, alarms, motion sensors and surveillance cameras to monitor and protect shelter premises; install physical barriers like doors windows fences etc. to restrict unauthorized access and institute visitor screening protocols or security patrols to deter criminal activity while increasing safety in shelter premises.

## Training Residents on Safety Procedures

Training residents on safety procedures is vital to raising awareness, being prepared, and taking proactive response in case of emergencies. Hold regular safety orientation sessions or workshops that educate shelter occupants on evacuation procedures, fire safety basics and first aid basics as well as personal safety tips. Provide hands-on demonstrations of equipment such as fire extinguishers, first aid kits and emergency exit and encourage residents to actively take part in emergency preparedness and response activities to empower themselves take ownership over their wellbeing and take charge.

## Lighting an Entire Shelter

Lighting up all areas of a shelter is paramount to creating a secure, safe, and pleasant atmosphere, especially at nighttime or low light conditions. Make sure all indoor and outdoor areas - such as hallways, stairwells, parking lots and entry points - are brightly illuminated without dark or shadowy areas; install energy-efficient fixtures (LED bulbs or motion activated lights) to increase visibility while deter criminal activity; periodically inspect and maintain lighting systems to replace burnt-out bulbs or repair damaged fixtures promptly.

## Maintaining Stockpiling Kits of Food

Stockpiling food, water and first aid supplies are key for being ready and resilient in times of emergencies or disaster. Prep food such as canned goods, dried fruit and granola bars with long shelf lives that can easily be prepared, store an ample supply of potable water in containers that don't leach odors such as containers with air filters to meet hydration needs of shelter occupants, as well as first aid kits stocked with bandages, medications, antiseptics, and emergency equipment as part of emergency plans.

## Maintain Hygiene and Sanitation Facilities

Maintaining sanitation facilities in shelter environments is vital to prevent infectious disease transmission and foster overall health and well-being. Clean and disinfect communal spaces, restrooms, showers and kitchen facilities on an ongoing basis in order to uphold hygiene standards; provide ample supplies of soap, hand sanitizer, toilet paper and personal hygiene products as needed by shelter occupants; ensure proper waste disposal practices such as garbage collection and management, pest control measures as well as proper waste disposal methods such as garbage collection as well as inspection of plumbing drainage and ventilation systems so as to detect water leaks as well as indoor air quality issues before occurring.

## DIY Project: Shelter Maintenance Kit: Keeping Clean and Safe in Emergency Shelters

**Materials:**

- Cleaning supplies (disinfectant wipes, multi-surface cleaner, bleach)
- Trash bags.
- Broom and dustpan
- Portable handwashing station or hand sanitizer
- Personal hygiene items (soap, shampoo, toothpaste, toothbrushes)
- Towels or washcloths
- **Laundry detergen**t
- Dishwashing soap and sponge
- Gloves
- First aid kit
- Fire extinguisher.
- Carbon monoxide detector (if using heating devices)
- Flashlights or lanterns with extra batteries
- Emergency contact list
- Emergency shelter guidelines or rules
- Waterproof tarp or ground cover
- Duct tape.
- Zip ties.
- Rope or paracord
- Portable toilet or waste disposal bags

**Procedure:**

- Starting by organizing all the materials listed above into a separate shelter maintenance kit. Store these in an easily transportable container such as a waterproof bag for optimal transportability and accessibility.
- Regular cleaning and sanitation in emergency shelters is vital to maintaining cleanliness and limiting germ proliferation. Utilize disinfectant wipes or multi-surface cleaner to clean frequently touched surfaces such as doorknobs, tables, and countertops with regularity.

- Proper trash disposal requires placing trash bags into designated bins outside the shelter and disposing of them at once.
- Sweep the floors regularly using a broom and dustpan to remove dirt, debris, and any contaminants, paying special attention to high traffic areas or areas prone to spills or other types of messiness.
- Maintain proper hygiene among residents at shelters by offering access to handwashing stations or hand sanitizer. Also make sure that soap, shampoo, toothpaste, and toothbrushes are readily available as part of their personal hygiene needs.
- Make a habit of regularly laundering towels, washcloths and other linens using laundry detergent and following manufacturer recommendations on washing and drying time. This ensures they remain both hygienic and clean for use by family members or for use by clients.
- After each use, clean dishes and utensils with dishwashing soap and sponge - using hot water when available and leaving items to air-dry before storing.
- Wear gloves whenever handling cleaning supplies or meeting potentially contaminated surfaces to protect yourself against germs and bacteria.
- Make sure to keep a first aid kit stocked with essential medical supplies and medications to treat minor injuries or illness, with regular inspection and replenishing, as necessary.
- Install fire extinguishers at strategic areas within your shelter and ensure all residents know how to use them during emergency fire situations.
- Install carbon monoxide detectors when using heating devices like portable heaters or stoves to monitor air quality and avoid carbon monoxide poisoning.
- Provide adequate lighting within the shelter with flashlights or lanterns and keep extra batteries on hand as required for replacement should any occur.
- Make an emergency contact list that includes phone numbers and addresses of local authorities, medical facilities, and any emergency services that might be helpful during an emergency. Post copies prominently within your shelter for easy access.
- Assign and communicate shelter guidelines or rules to residents to promote cleanliness, safety, and respectful behavior within the shelter environment.
- Assemble waterproof tarps or groundcover to provide shelter floor protection against moisture, dirt, and debris. Secure them as required using duct tape, zip ties, rope, etc.
- Check your shelter regularly to detect any signs of wear or damage and take timely action to repair leaks, holes, or structural problems to preserve its integrity and keep everyone safe.
- Provide portable toilets or waste disposal bags so residents can use these facilities when restrooms are unavailable and decide to dispose of their waste in designated areas outside the shelter.

# BOOK 20: GARDENING FOR PREPPERS: ESSENTIAL GUIDE

Selecting an ideal location for your survival garden is vitally important to its success. Find a spot with ample sunlight during the day (ideally at least six hours of direct sunshine) as this will allow your plants to get enough sustenance from sunlight. In addition, ensure adequate drainage as wet soil can lead to root rot or other issues; proximity to an irrigation source such as a hose or rainwater collection system should also be considered; choose something accessible so tending and harvesting produce can easily take place.

## Understanding Soil Composition and pH Levels

Understanding your garden area's soil composition and pH levels is critical for optimal plant development. Conduct soil tests to ascertain its texture (sand, silt, or clay), texture type and pH range - most vegetables prefer slightly acidic to neutral soil within a pH range between 6.0 to 7.0; amend as necessary with compost, aged manure, or mulch as organic material to increase nutrients while improving moisture retention and structure.

## Selecting Appropriate Plants Based on Climate and Soil Conditions

Selecting plants adapted to both climate and soil type is crucial in creating an effective survival garden. When researching varieties to plant in your local climate, consider factors like temperature range, frost dates, rainfall amounts and the length of day length; then match these choices to soil types such as sandy loam or clay and consider disease resistance, pest tolerance as well as growth habits when making decisions regarding plant selections.

## Planning Your Garden Layout to Achie Optimum Yield

Careful consideration can boost yield and efficiency when planning your garden layout. Group plants with similar water, sunlight and nutrient needs together for easy care and maintenance; utilize vertical space by attaching climbing plants such as tomatoes or cucumbers onto trellises/stakes/cages for climbing purposes like tomatoes/cucumbers/beans to them for growing at height; implement succession planting to stagger harvests while prolonging growing seasons by replanting areas as crops are harvested; leave sufficient gaps between rows/plants for air circulation/weed control/ease of access

## Companion Planting

Companion planting is a gardening strategy which utilizes multiple species of plants together to meet each other's needs. Some produce natural compounds which repel pests or attract beneficial insects while others fix nitrogen into the soil or offer shade and support. Research companion planting combinations which complement one another in terms of growth habits and pest resistance - for instance planting marigolds near tomatoes may deter nematodes while placing basil near tomatoes can enhance flavor while deterring pests - then experiment to see which strategy best serves your garden!

## Seed Saving and Preservation Techniques

Seed saving and preservation techniques are key for providing your survival garden with sustainable supplies of plants. Choose healthy mature plants with desirable traits for seed saving; allow seeds to fully mature on their plants before harvesting for cleaning and drying; store in airtight containers like glass jars or envelopes labeled with plant variety name and harvest date until ready for storage; rotate seed stocks regularly to maintain viability without loss due to deterioration or degradation.

## Building Raised Beds or Container Gardens for Limited Space

Building raised beds or container gardens can provide an efficient means of gardening in limited spaces. When creating raised beds using untreated lumber, bricks, or any durable material suitable for construction you should create defined growing spaces with improved drainage, soil structure, and drainage; fill them with topsoil, compost and organic matter mixture for optimal plant growth; use containers like pots buckets grow bags etc. to cultivate vegetables herbs flowers on patios balconies rooftops ensuring containers have adequate drainage holes as well as support to promote proper root development in order to be most successful when cultivating vegetables herbs flowers vegetables herbs flowers grown this way!

## Watering Techniques to Achie Optimal Plant Health

Watering techniques are critical in survival gardens to ensure optimal plant health and survival. Regular, deep and infrequent irrigation will encourage deep root development and drought tolerance; morning irrigation helps limit evaporation while decreasing fungal disease risks; drip irrigation systems (drippers or soaker hoses), soaker hoses or watering cans) deliver water directly to plant bases reducing waste water usage; mulch garden beds using organic materials like straw, wood chips or leaves in order to retain soil moisture and inhibit weeds growth in beds - just two things that ensure optimal plant health!

## Fertilization Options Including Organic and Synthetic Solutions

Fertilizing methods provide essential nutrients needed for plant health and productivity in survival gardens. Choose between organic options like compost, aged manure, worm castings or organic fertilizers from plant or animal sources; apply according to package directions or incorporate before planting; alternatively synthetic nitrogen-phosphorus-potassium fertilizers may address specific deficiencies or boost plant growth more effectively by following recommended rates and applications timing to avoid overfertilization and runoff of nutrients.

## Survival Gardens Provide an Ideal Platform

Pest and disease management is essential to crop preservation and garden productivity. Implement integrated Pest Management (IPM) strategies to minimize chemical pesticide reliance and promote ecological balance, monitoring plants regularly for signs of pests, diseases, or nutritional deficiencies before using physical barriers such as row covers or netting to exclude these dangers or protect plants from potential damage from them. Practice crop rotation, companion planting or intercropping as ways of disrupting pest cycles while strengthening resilience while natural predators such as beneficial insects can effectively manage populations as biological controls.

## Season Extension Solutions

Season extension methods can extend the growing season and protect crops in survival gardens by protecting from harsh weather conditions. Construct cold frames, hoop houses or row covers out of lightweight materials like PVC pipes, plastic sheeting, or recycled materials to capture maximum sunlight and provide enough ventilation - use floating row covers or cloches on individual plants if frost or cold snaps appear while monitoring temperatures inside season extension structures and ventilating as necessary to avoid overheating or humidity build-up.

## Vertical Gardening for Space Efficiency

Vertical gardening can be an efficient and space-efficient method of expanding survival gardens in small or urban settings, particularly by growing vining crops like cucumbers, beans and peas on trellises, arbors, or fences to save ground space while improving air circulation. Vertical structures like pallets shelves or hanging baskets may be utilized to grow herbs salad greens strawberries vertically using stakes strings supports or stakes and train these vines upward using stakes strings supports; employ vertical gardening techniques in limited garden areas for greater yield and diversity!

## Permaculture Principles for Sustainable Gardening

Permaculture offers a holistic approach to gardening that takes advantage of nature to foster productive ecosystems that support life. Implement the following permaculture principles into your gardening design and practices for maximum sustainability:

- Observation and Interaction: Analyze and understand the natural patterns, processes, and interactions within your garden ecosystem to design and manage it effectively in terms of productivity and sustainability.
- Integrate Diversity: For an eco-friendly garden ecosystem that thrives and remains resilient over time, embrace diversity among plant species, habitats, and wildlife by including crops, trees, shrubs, and flowers that support biodiversity, enhance soil health, and deter pests in your planting plan.
- Use Resources Wisely: Utilize natural resources efficiently and sustainably such as sunlight, water, soil, and organic matter - such as sunlight, water, soil, and organic matter - such as sunlight, water, composting, or mulching techniques to conserve resources while minimizing waste. Water saving measures, composting or mulching methods may help.
- Apply Design Principles: Utilize permaculture principles such as stacking functions, zoning, and edge effect in designing your garden to achieve maximum space, productivity, and functionality. Plan for efficient usage of space, energy, and resources to produce a harmonious yet productive garden layout.
- Promote Natural Processes: Mimic and enhance natural processes like nutrient cycling, soil regeneration and pest control to maintain ecosystem balance and resilience. Implement practices such as companion planting, crop rotation and biological pest control which support these natural systems while decreasing dependence on external inputs.
- Accept Change: Recognize and adapt to changes within your garden ecosystem over time, such as seasonal variations, climate fluctuations and ecological succession. Permaculture gardening's core principles emphasize being flexible and resilient to adapt quickly to ever-evolving circumstances.

By following permaculture principles such as observation, diversity, resource efficiency, design principles, natural processes, and adaptability you can build a resilient garden ecosystem in harmony with nature.

## Mulching and Weed Control Strategies

Mulching is an integral practice in gardening that conserves soil moisture, suppresses weeds, moderates soil temperature, and enhances overall soil health. Consider these effective mulching and weed control strategies to maintain an efficient garden:

- Select Organic Mulch Materials: Organic mulch materials like straw, hay, leaves, wood chips or grass clippings make great organic covers to cover soil surfaces around plants and provide essential nutrients while decomposing in time to improve soil structure over time.
- Apply Mulch: Spread an even layer of mulch evenly around plants, leaving an air gap at each stem edge for moisture accumulation and potential plant rot to escape. Aim for depths between two to four inches to effectively suppress weeds while conserving soil moisture levels.
- Weed Barrier: Use cardboard, newspaper, or landscape fabric as a weed barrier under mulch to further suppress weed growth and stop seeds from germinating. Make sure the barrier allows water and air penetration while at the same time blocking sunlight to effectively inhibit further weed development.
- Regular Mulching: Maintain the effectiveness and appearance of mulch by replenishing its layers regularly to preserve its effectiveness and appearance. Organic mulch decomposes over time; add additional layers as it decomposes to replenish soil nutrients, suppress weeds, and maintain soil moisture levels.
- Practice No-Till Gardening: Reduce soil disturbance and compaction by engaging in no-till gardening techniques. Avoid digging or cultivating soil excessively as this could disrupt its structure, expose weed seeds, and promote their proliferation.
- Hand Weeding: Monitor your garden regularly for signs of weed growth and remove weeds by hand before they take root and spread further. Hand weeding can provide effective yet eco-friendly weed control without resorting to harsh chemical herbicides or mechanical cultivation processes.

Implementing mulching and weed control strategies such as selecting organic materials for mulch application, applying it properly, using weed barriers where applicable, mulching regularly as part of no-till gardening and hand weeding can ensure you maintain an attractive garden ecosystem free from unwanted weeds.

## Integrating Medicinal Herbs and Edible Flowers into Your Garden

Integrating medicinal herbs and edible flowers into your garden not only adds visual interest and diversity but can provide important resources for culinary, medicinal, and therapeutic uses. Follow these tips on incorporating them:

- Select Appropriate Varieties: It is important to choose medicinal and edible flowers which suit the climate, soil conditions and garden space you have available to you. Research plant varieties which thrive well within your region while offering culinary or medical benefits that provide what is desired in terms of culinary or medicinal uses.
- Plan Garden Layout: Create your garden layout so it includes specific areas or beds dedicated for cultivating medicinal and edible flowers, considering elements like sunlight exposure, soil quality and water availability when selecting plant placement locations.
- Companion Planting: Companion planting involves interplanting medicinal and edible flowers with vegetables, fruits, or ornamental plants to increase space efficiency, enhance biodiversity and deter pests. Select companion plants which complement one another in terms of growth patterns, nutrition needs and pest resistance.

- Provide Appropriate Care: Ensure the proper care of medicinal herbs and edible flowers by providing adequate water, sunlight, and soil nutrients. Mulching around plants helps conserve soil moisture, suppress weeds, and maintain soil temperature.
- Harvesting and Processing: At their most fresh and potent, medicinal, and edible flowers should be collected at their peak of freshness for culinary or medical use. Dried herbs can be stored or preserved, while fresh flowers and leaves can be added directly into teas, salads, herbal remedies, or decorative arrangements for use as fresh additions or decorative accessories.
- Examine Uses and Benefits: To take full advantage of medicinal herbs and edible flowers' full potential, investigate all their culinary, medicinal, therapeutic uses and benefits in terms of culinary use, medical use, and therapeutic benefits as well as preparation methods, recipes, and applications to uncover all their varieties of flavors while enjoying their health advantages. Experimentation will reveal all flavors as well as health advantages that exist with their use.
- Integrating medicinal and edible flowers into your garden will bring beauty, diversity and functionality while offering culinary pleasure and health benefits for you and your loved ones.
- Promoting pollinator-friendly gardening practices is crucial to supporting biodiversity, increasing food production, and conserving native pollinator species such as bees, butterflies, and hummingbirds. Implement these practices in your garden ecosystem for maximum pollinator friendliness:
- **Plant Native Flowers: When selecting native flowering plants to offer nectar and pollen sources for local pollinators species throughout their growing season. Look for varieties in shapes, colors, and bloom time to attract as wide an array of pollinator species as possible.**
- Provide Shelter and Nesting Sites: Establish habitat and nesting sites for pollinators by including features like brush piles, rockeries, log piles and nesting boxes in your garden landscape design. Wind and rain protection should also be ensured as well as not using pesticides that might harm pollinators' populations.
- Avoid Hybridized Plants: Cultivar plants do not typically offer native pollinators the nectar and pollen resources that they rely upon, making heirloom or open pollinated varieties the more effective choices when selecting pollinator-friendly varieties of plants.
- Include Host Plants: Include host plants in your garden to assist the life cycles of butterflies, moths, and other pollinators insects. Host plants provide food and habitat for caterpillars and larvae feeding off these host plants and increasing pollinator populations overall.
- Provide Water Sources: Provide shallow sources, like birdbaths, puddling areas, or small ponds for pollinators to drink and bathe from. Make sure these sources remain clean and refreshed regularly to prevent stagnation and mosquito breeding.
- Avoid Chemicals: To help safeguard pollinators and conserve pollen sources from potential harm caused by harmful toxins and chemicals used as garden fertilizers and pesticides. Instead use natural and organic methods of control such as handpicking pests manually, companion planting arrangements and beneficial insect releases as alternatives for protecting pollinators health in your garden.
- Maintain Continuous Bloom: To provide pollinators with continuous nectar and pollen sources throughout the growing season, select flowering plants that bloom at various points throughout the

growing season, from early spring bloomers through summer flowerers, then fall-bloomers. Doing this ensures year-round support of pollinator populations.

By adopting pollinator-friendly gardening practices such as planting native flowers, providing shelter, and nesting sites, avoiding hybridized plants, providing water sources without chemicals use, maintaining continuous bloom, you can foster an ecosystem which supports pollinator health and biodiversity.

Harvesting and Storing Produce for Long-Term Use Storing produce for future consumption allows you to take full advantage of your garden throughout the year while decreasing food waste. Follow these tips when harvesting and storing produce to maintain freshness, flavor, and nutritional value:

- Harvest at Peak Ripeness: For optimal taste, texture, and nutritional content in fruits and vegetables, harvest at their optimal point of ripeness using visual cues such as color, size, and texture to determine when to pick. Also harvest them early morning when sugar content is highest for maximum impact!
- Be Gentle with Produce: Handling produce with care can prevent it from being bruised, crushed, or otherwise compromised in its integrity. Use clean tools like scissors and pruners when harvesting produce to harvest effectively while pulling or twisting stems excessively may damage them further.
- Remove Excess Foliage: To minimize moisture loss and extend shelf life of harvested produce, it's a good practice to trim excess foliage off harvested products before they have reached shelf storage. You should leave some stems attached as necessary to preserve freshness and flavor.
- Clean and Dry: Once harvest is completed, produce must be thoroughly washed in cold running water until cold running water runs clear before pat drying with either a clean towel or paper towel to eliminate dirt, debris and pathogens that might contribute to spoilage. Rinsing fruits and vegetables under this method also provides cold running water which aids drying before storage.
- Select Appropriate Storage Solutions: To preserve freshness and prevent spoilage, store produce in containers that allow airflow. Breathable containers like perforated plastic bags, mesh bags or vegetable storage bins help encourage airflow while simultaneously decreasing moisture build-up.
- Control Temperature and Humidity: For optimal storage conditions, place produce in cool and humid environments like root cellars, basements, or refrigerator crisper drawers to slow ripening while also avoiding wilting or decay. Monitor both temperatures and humidity regularly to guarantee ideal storage conditions.
- Avoid Ethylene Exposure: To limit exposure to ethylene-producing fruits like apples, bananas, and tomatoes which emit this gas, leafy greens, herbs, and berries should be stored away from these items to reduce premature ripening and spoilage.

Rotate and Use First-in, First-Out (FIFO) Methodology when Storing Produce: Rotation and First-in, First-Out inventory management should be practiced when storing produce to ensure older items are consumed first before more recent ones come along. Label containers with harvest dates or expiry dates to monitor freshness levels more effectively while decreasing food waste.

By following these harvesting and storage practices, you can maximize the shelf life and quality of garden produce and enjoy its flavors throughout the year.

# DIY Project: Survival Garden: Growing Your Green Thumb for Self-Sufficiency

**Materials:**

- Seeds of various vegetables, fruits, and herbs (choose varieties suited to your climate and growing conditions)
- Soil amendments (compost, manure, peat moss)
- Gardening tools (shovel, rake, hoe, trowel, watering can)
- Raised garden beds or containers.
- Garden fencing or netting to protect from pests.
- Mulch (straw, wood chips, or grass clippings)
- Watering system (hose, drip irrigation, or watering wand)
- Sunscreen and gardening gloves
- Garden markers or labels.
- Pruning shears or scissors
- Insect repellent
- Plant supports (stakes, cages, or trellises)
- Planting calendar or guide
- **Harvesting baskets or containers**
- **Plant nutrients (fertilizers or compost tea)**
- Pest control products (organic options preferred)
- Cold frames or row covers for extending the growing season.
- Garden journal or notebook.
- Plant propagation materials (rooting hormone, propagation trays)
- Emergency backup seeds and supplies

**Procedure:**

- Start by finding an ideal spot for your survival garden. Aim to select an area which gets at least 6-8 hours of sun each day and has access to water sources that allow for irrigation purposes.
- Prepare the soil by loosening it using a shovel or garden fork and adding soil amendments such as compost, manure or peat moss as needed to improve fertility and drainage.
- Consider how each plant will mature before planning your garden layout; take note of size requirements for matured specimens as well as spacing issues when setting out your design. Consider raised garden beds or containers if limited space or poor soil quality exist in your space, to help maximize yield from each space you can cover with plants.
- Choose vegetables, fruit and herbs that thrive in your climate and growing conditions - especially heirloom or open pollinated varieties for long-term sustainability and seed saving purposes.
- Protect your garden from pests and wildlife by installing garden fencing or netting around its perimeter, employing organic methods such as companion planting, natural predators, or homemade insect repellents as effective safeguards.
- Mulch the area around your plants with straw, wood chips or grass clippings to reduce weeds, preserve moisture levels and regulate soil temperatures. This will also keep weeds at bay!

- Establish a watering plan to provide your plants with sufficient hydration during periods of drought or extreme heat, including periods when rainfall may not meet demand. Drip irrigation or soaker hoses may help save water by conserving resources while decreasing evaporation rates.
- Protect yourself from harmful UV rays with sunscreen, and wear gardening gloves to avoid blisters or injuries while working in your garden.
- Garden markers or labels allow you to keep an eye on their progress throughout the growing season and identify which plants they belong to.
- Trim and train your plants as necessary to promote healthy development and maximize yields. Use pruning shears or scissors to cut away dead or damaged foliage before supporting tall vines with stakes, cages or trellises for maximum support and yields.
- Consistently monitor your garden for signs of pests, diseases, or nutritional deficiency - any signs that arise should be resolved swiftly to reduce any chance of further spreading and ruining harvests.
- As part of your crop planting and harvesting schedule, follow a planting calendar or guide. It can help ensure crops are being planted at optimal times in your region.
- Harvest baskets or containers allow you to efficiently collect fruits, vegetables, and herbs as soon as they reach maturity - this ensures continuous production throughout the growing season. Harvest regularly to maximize production efficiency.
- Provide your plants with additional nutrition as required by using organic or compost tea fertilizers to fertilize.
- Keep pests under control with natural, eco-friendly methods like handpicking insects, applying homemade repellents, or adding beneficial insects into your garden.
- Extend the growing season in cooler environments by protecting plants with cold frames or row covers to shield from frost, cold temperatures, or both.
- Maintain a garden journal or notebook to document all your observations, successes, and failures during each growing season. Use this data to improve your gardening abilities and plan future crops.
- Propagate your favorite plants by cuttings or division of mature ones, using rooting hormone and propagation trays for proper root development and to establish new plants.
- Store emergency backup seeds and supplies in a cool, dry location so you'll always have access to food and seeds during future growing seasons.

# CONCLUSION

As we end of The Prepper's Survival Guide: 20-in-1, it is crucial that we reflect upon what has transpired over this journey together. Over these pages we have covered key principles related to emergency preparedness, self-reliance, and resilience against adverse situations; from basic survival kit building techniques and off grid living strategies through to advanced strategies designed to protect self-defense tactics - everything needed for you to thrive under any condition has been covered here.

But our journey doesn't end here: preparedness is not a destination; rather it is an ongoing commitment to learning, adaptation and growth. So, as you set forth on your own path toward long-term resilience remember that adaptability, improvisation, and perseverance will always remain key components.

Be alert, aware and connected with your community during times of crises. Our greatest strength comes from coming together as one and supporting one another - whether facing natural disaster, social collapse, or personal emergencies, know that you aren't alone - reach out to neighbors, friends and fellow preppers for guidance, assistance, and camaraderie.

Keep hope alive; never lose it! Even during difficult times, our resilience, resourcefulness, and determination help carry us forward. By adopting preparedness principles such as self-reliance you not only protect yourself and loved ones from disaster; you also contribute towards building a brighter and more resilient future for us all.

As you step off this guide and embark upon the next leg of your journey, do so with optimism, courage, and conviction. Though the future may remain unpredictable, with knowledge, skills, and mindset of a prepper behind you you're better equipped than ever to tackle whatever obstacles lie in its way.

Thank you for joining us on this adventure, may your path be filled with resilience, strength, and hope!

# BONUS

Bonus 1

Mistakes Preppers Make When Storing Water

Bonus 2

15 Items Every Prepper Should Hoard

Bonus 3

Foods to Keep in your prepper pantry

Bonus 4

The First $1000 a New Prepper Should Spend

Bonus 5

Best Ways To SECURE - DEFEND Your Home & PREPPING Supplies

Made in the USA
Las Vegas, NV
14 May 2024

89906115R00079